AF837040

Managing Solar Energy

Fundamental Science to Practical Systems

in plain language

OMAR S. CHEEMA

ISBN 978-1-0369-3471-2

November 2025

Revised June 2026

Dedicated to Jerry P. Gollub, Haverford College

LEGAL DISCLAIMER

CONTENT

Foreword

This book is deliberately written off the cuff to serve as a practical guide for the planning, building, and management of solar photovoltaic (PV) systems, and crafting policies for optimal application including innovative electric grid integration. We assume readers have no prior detailed knowledge of the related science or business, but took some physics and mathematics in school – enough to feel comfortable with algebraic notation and reading equations.

It is not meant as a definitive technology handbook or science textbook. The primary intent is to help develop scientific and commercial intuition about the subject, provide valuable insights into the principles of how PV systems work, their full potential in modern electric power, so as to more meaningfully and effectively engage in this business from a professional practitioner's perspective.

A multifaceted holistic perspective, comprehending basic aspects of electric power systems, is essential for a well-educated PV professional. Technical detail is laid out first, and then the economic analysis that has to be based upon it for a full techno-economic picture. It is written in the global public interest, to contribute to professional learning in solar electricity, but also for a broader audience to objectively understand the topic, rather than another marketing piece for solar energy or peer-review publication exercise.

You can make up your mind, after reading what follows, whether it serves the purpose. We welcome feedback for improvements, any corrections or clarifications, to refine the book.

Ch1. Sunlight and Electricity

"It is important to realise that in physics today, no one knows what energy is."

– Richard Feynman[1]

A solar PV system directly converts sunlight into electricity. Before detailing the PV system, we have to understand sunlight and electricity i.e., outline what is being converted to what. The first concept to grasp is that of energy itself. Both sunlight and electricity are forms of energy. Rather than trying to come up with some abstract definition of energy, it is easier to dive into how energy is found and utilised. Various forms of energy can be defined, identified, measured, and harnessed. For practical purposes, the concept of energy is most conveniently understood in terms of its characteristics. The primary characteristic of energy to understand is that it is conserved in total i.e. the law of conservation of total energy over time. This conservation law applies to the whole universe and any part of it that is not allowed to exchange energy with other parts. We humans do not create or destroy any energy but merely convert energy from one form to another, transferring energy from one part of the universe or any section of it to another. We are involved in the business of energy conversion and transfer to suit our needs, for whatever we plan to do with it.

Nature has been busy in converting energy from one form to another and transferring it from one part of the universe to another, since forever. Otherwise, we would not exist.

Our instinct for survival taught us how to further convert and transfer energy, going beyond what the prevalent processes accomplished. All living species are naturally programmed to play this game. As our scientific knowledge of nature progressed, we learnt how to develop the game in more sophisticated ways.

Forms of Energy

The fundamental forms of energy that we have discovered include thermal, gravitational, mechanical, chemical, nuclear, and electromagnetic. Electromagnetic energy, of particular interest to us, is electrical and magnetic energy coupled together - joined at the hip so to speak. These forms of energy are governed by the laws of nature. The energy conversion and transfer game cannot circumvent or beat these natural laws.

Thermal

This form of energy is arguably the most important to understand, as it is the ultimate limiting factor in energy conversion and transfer. Thermal energy arises from the haphazard internal motion of particles that make up the body of a physical object. The energy of motion, a concept we are all familiar with in our everyday lives, is called kinetic energy. Constituent particles of any type of physical matter, whether a gaseous, liquid, or solid body, have some degree of random motion. The body as a whole might be stationary, but it has internal kinetic energy from this random motion of its constituents. Internal kinetic energy does not make the body move as a whole, but results in what is called the temperature of the body. Temperature is thus a measure of thermal energy. A hotter temperature indicates greater random motion inside the body. If the random motion of the constituent particles is reduced to zero, the body reaches absolute zero temperature. This extreme condition can be thought of as the definition of absolute zero temperature; the temperature scale used for measuring it is called the Kelvin scale with zero degrees Kelvin, or 0 Kelvins as is now preferred, = minus 273.15 degrees Centigrade or minus 459.67 degrees Fahrenheit. It is practically impossible to achieve such a temperature, at least for the purpose of our planning.

Whenever two bodies come in contact, there is a transfer of thermal energy. The hotter temperature body transfers net thermal energy to the other. The process of thermal energy transfer is defined as heat (heat flux being the thermal energy transfer per unit area per unit time). Different physical processes of conduction, convection, or radiation bring about the thermal energy transfer till the internal random motion inside the two bodies is balanced.

Gravitational

Gravitational attraction exists between any two bodies in the universe. It's critical to how the entire universe hangs together. Our interest is in the gravitational attraction of bodies located on the planet we live upon - the gravitational energy potential between any body and our planet. Potential specifically refers to a configuration of energy capacity stored in that form. Gravitational energy is proportional to the mass of the body (the size and number of molecules constituting the body), and the height of the position of the body from a surface below it. A change in the height of the body converts its gravitational energy into kinetic energy. Spatial positions of bodies, with respect to each other, are mediated by their gravitational attraction, e.g., the distance of the Earth from the Sun in its elliptical orbit. Gravitational energy is a manifestation of attraction between the masses of different bodies. Mass only results in attraction. The greater the masses, the greater the attraction between them (a product of the two masses), and the higher their gravitational energy potential.

Mechanical

Mechanical energy is classified as potential or kinetic. Potential energy is stored in the relative positions of mass. A compressed spring stores potential energy that is also referred to as strain energy. Releasing the spring converts its potential energy to kinetic energy.

Kinetic energy comes from the external motion of the body as a whole. When a solid body moves against another, kinetic energy is converted to thermal through the well-known but complicated physical phenomenon of surface friction. Friction, measured in viscosity, is also present in liquids and gases between the internal movement of the constituent particles.

When a moving body collides with another moving or stationary body, kinetic energy is transferred. In a perfectly elastic collision, total kinetic energy is conserved. Practically, some of the kinetic energy is converted to sound energy and thermal (we are not considering the elastic collisions of subatomic particles of matter or anti-matter here, like electrons, protons, positrons). Sound is ordered longitudinal motion of the constituent particles, as opposed to random thermal motion.

In a perfectly inelastic collision, the colliding bodies come together for a maximum transfer of kinetic energy to other forms. Kinetic energy in inelastic collisions can be converted to sound, heat, light, chemical energy of broken molecular bonds, gravitational potential energy in the height gained by colliding bodies, or the strain energy of deformation from shattered bodies. Friction between the surfaces of colliding bodies can contribute to lethal energy release, as in a bullet penetrating the human body.

Changes in kinetic and or potential energy are referred to as mechanical work done or work done for short. Kinetic energy imparts the physical property of momentum to a body. Momentum is conserved unless the body interacts with another source of energy. Force is defined as the rate of change of momentum. Mechanical work done is the integral of the force over the distance it moves an object.

Chemical

Chemical energy is contained in the bonds formed by electrons circling their respective atoms to constitute molecules - electronic combinations representing different types of chemical bonds. Energy is released or absorbed when molecules interact to make a different set in what is called a chemical reaction. Biological energy, also called biochemical energy or bioenergy for short, is a subset for chemical reactions within living species or renewable physical matter produced by the living species.

Nuclear

Nuclear energy is trapped in the nucleus of an atom: the energy required to hold the nucleus together. Nuclear reactions release or absorb the binding energy by destruction or creation of new nuclei – the processes of fission and fusion. Fission splits the nucleus of a bigger atom into smaller atomic nuclei. Fusion combines the nuclei of smaller atoms to make a bigger atom. The energy output is heat and electromagnetic radiation.

Electromagnetic

Charged particles of electrons and particles of light called photons are the main protagonists of electromagnetic energy. Negative charge of an electron creates an electric field: a field of electrically charged energy. By a “field”, we mean a spatial and temporal map of energy gradients, a dynamic distribution in the manifold of spacetime. When an electron moves, a field of magnetic energy is generated (a relativistic effect explained by the special relativity theory of spacetime). Electric and magnetic fields synchronously couple at the speed of light to produce electromagnetic energy. Unobservable virtual photons (particles of light) are theoretically conceptualized to mediate the interactions of electromagnetic fields. Real observable photons, travelling at an invariant velocity "c" in vacuum, carry electromagnetic energy. The velocity "c" is a fundamental universal constant. A photon bears the smallest measurable quantum (i.e., permitted quantity) of an electromagnetic energy spectrum. Light comes in a variety of permitted energies that different photons can carry.

Sunlight

Photons are constantly being emitted by all physical matter with discrete levels of energy. Permitted energy levels and the quantities of photons for each level, depend on the chemical composition and temperature of the physical body. The transfer of electromagnetic energy by photons is called radiation.

The Sun has been regularly radiating electromagnetic energy to Earth for a few billion years. A nuclear fusion reactor at its core, created by immense gravitational energy, sustains a plasma with a photosphere at ~6000K emitting electromagnetic energy in close to what is called the blackbody spectrum (distribution of energies radiated from a perfect radiator and absorber of electromagnetic energy). Electromagnetic energy is emitted in an isotropic manner, i.e., uniformly in all directions. About $1.07 \cdot 10^{11}$ TWh per second from the Sun.[2] 1TWh = one billion kWh, and 1kWh = 3.6 million Joules, a Joule is the SI unit of energy (we will primarily use Wh, kWh, MWh, TWh for energy units in this guide). For power, the rate of transferring energy, we will use the units of watts = 1J/s. At this point, the distinction between power and energy is a useful reminder. Power is the rate of transfer of energy per unit time. Mathematically, Power, $P = dE/dt$, where E is the energy being transferred. Conversely, $E = \int P\, dt$, where the integration is over the finite period of the energy transfer. Power flux is the energy being transferred per unit time per unit area, in units of W/m^2.

High power does not necessarily mean much energy has been transferred, if the transfer time was a short period. Conversely, a large energy transfer does not mean high power usage if the time taken for it was very long. The relationship of energy to power is analogous to that between distance covered and the speed of travelling that distance.

Due to the isotropic nature of electromagnetic energy transfer by radiation, the amount of energy emanating from the Sun spreads over a spherical area of $4\pi r^2$. Isotropic reduction of energy transfer with distance is known as the inverse square law. By the time it reaches the upper atmosphere of the Earth, "r" the average distance between the Sun and Earth of ~150 billion meters,[3] the receipt is $1.07 \cdot 10^{11}$ TWh/s divided by $4\pi \cdot 150 \cdot 150\ 10^{18}$m giving ~0.38milli-kWh/s per m^2 or equivalently a solar power flux of 1361W/m^2.

Solar power flux comprises photons covering a wide range of energies - plotted in the solar spectrum shown in Figure 1:

Figure 1: solar spectrum at the top of the Earth's atmosphere

https://commons.wikimedia.org/wiki/File:Solar_Spectrum.png

- Robert A. Rhode (October 2008)

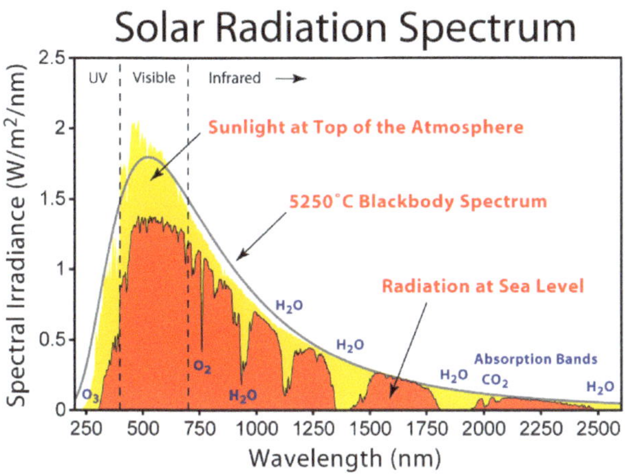

A total of around $3.6 \cdot 10^{21}$ photons per m^2 strike the top of the Earth's atmosphere - photons at the peak of the spectrum have an energy of 2.48eV (an electron-volt "eV" is yet another unit of energy used for very small amounts of it, $\sim 2.25 \cdot 10^{25}$eV = 1kWh). Photons radiate in an electromagnetic wave - coupled electric and magnetic fields sinusoidally oscillating in orthogonal directions to each other - so each photon energy can be assigned a corresponding wavelength. The equation relating photon energy to its wavelength is E = hc/λ where "E" is its energy, "h" a most fundamental physical constant of the quantum mechanics Uncertainty Principle, c" the velocity of a photon in vacuum, and "λ" the wavelength. The Uncertainty Principle is about how nature limits what information we can know about position and momentum or energy and time of a particle simultaneously. A photon of 2.48eV energy has a wavelength of 500nm (nm is nanometres: a billionth of a meter), showing a green colour to the normal human eye.

Since the space between the Sun and Earth is largely empty, there is not much interaction of the photons with any physical matter along the way. The photons are travelling at their velocity "c" in vacuum, reaching the upper atmosphere in around eight and a half minutes. They then have to pass through a multi-layered film of gases and clouds comprising the atmosphere to reach the terrestrial surface of the planet. The energy supply of these photons is responsible for all energy stored on Earth, with the exception of the planet's inner core internal energy, and its daily replenishment. Sunlight broadly refers to this photonic energy supply.

How many photons make it through to the Earth's terrestrial surface, and with what energies, depends on their interaction with the constituents of atmospheric layers: the phenomena of scattering, refraction, absorption, and re-emission. The Earth's weak magnetic field deflects some of the harmful charged particles coming at us from the Sun, and the upper atmospheric layers filter dangerous ultraviolet (UV) radiation. The Earth is a rotating near-sphere with a total surface area of $4\pi r^2$. It intercepts solar radiation as a flat disc facing the Sun with a cross-sectional area of πr^2. The total surface area for daily solar distribution is $4\pi r^2$. For a solar flux of ~1361W/m^2, the global daily average of sunlight distributed over the entire surface is $(1361W/m^2) \cdot (\pi r^2)/4\pi r^2 =$ ~340W/m^2. Roughly a third is rejected back to space, leaving ~230W/m^2 of sunlight available for terrestrial energy conversion.[4] The 230W/m^2 takes care of all photosynthesis on the planet, the chemical energy demand of bacteria, thermal energy for fungi and other living species including underwater creatures. It sets up the planetary heat engine for wind currents, seasonal weather patterns, and over millions of years the bacterial anaerobic digestion, catagenesis, of fossil fuels from dead organic matter. It also drives the hydrologic cycle for renewable freshwater resources on the planet. Due to the thick insulation of the Earth's mantle, only a small fraction of the heat from its core dynamo and radioactive decay reaches the surface: ~60-100mW/m^2, some of it released through volcanoes and natural hydrothermal systems.[5] A trivial amount compared to sunlight. The core dynamo has a critical protective role in maintaining the Earth's weak magnetic field, but it is sunlight that dominates the global energy budget.

We shall later see that only a fraction of the total available sunlight is suitable for absorption by a single semiconductor material or multiple semiconductor layers. A large number of infrared (IR) photons, and a smaller number in the ultraviolet (UV) range, are unfit for the job. In addition, there is a physical trade-off between the number of photons that can be captured and their selected energies.

Electricity

Electricity is electromagnetic energy guided in an electrical circuit to meet the demand of an energy consuming element called the load. Electrical circuits are closed loops for precisely transferring this energy from the source to the load. Depending upon the design of the circuit wires/cables, whether the conductors are insulated or not, electromagnetic energy either externally flows along the surface of or is internally confined to the electrical cables. Energy flows around the metal conductor, not through the metal, to the load. Some is lost as thermal energy dissipated in the conductors. Surface charges on the metal guide the flow.[6,7] Steady-state or oscillating electrical (E) and magnetic fields (B) are established for the energy transfer. Voltage and current are set up in the circuit, with associated electrical and magnetic fields respectively. Charge carriers carry electrical current in the direction of the electric field (the convention is to show positive electrical current in the opposite direction to electron movement). The product of voltage and current gives the electric power of the circuit, i.e., the rate of transferring electrical energy from the source to the load. Steady-state electrical and magnetic fields drive Direct Current (DC), while oscillating electrical and magnetic fields produce Alternating Current (AC). Whether DC or AC is used, the electromagnetic energy transfer occurs in the direction of the flux vector S, a cross-product of the E and B fields: $S = (1/\mu) \cdot (E \times B)$, in units of W/m^2, where "μ" is the permeability of the medium for the energy transfer, and "x" denotes a vector cross-product.[8] The energy transfer vector S points in the same direction for both DC and AC, from the source to the load, given by the cross-product between E and B.

If the electrical conductors in the circuit are uninsulated, the electromagnetic power transfer proceeds at close to the velocity of light in vacuum. For insulated cables, the velocity is somewhat reduced by the dielectric insulation of the conductor. In either case, electricity can be transferred across the world with lower energy losses than other forms of energy harnessed by humans. It can also be routed and shared in sophisticated patterns, with the voltage and current manipulated for all sorts of power flows.

As with all energy conversion and transfer processes, a key consideration is the minimization of heat dissipation i.e., a by-product of thermal energy wasted as heat that cannot be recovered. To understand this common problem in more detail, we need to go through the fundamental rules of nature governing energy conversions and transfers.

There is a deeper meaning to energy in the fundamental physics of relativistic spacetime and quantum mechanics. Energy is physical information that manifests itself in the various forms we have discussed. The theory of special relativity describes the invariance of energy squared in different frames of reference. Quantum mechanics highlights how the measurement of energy and time are interrelated. Are there forms of energy that we have not yet discovered and understood? Perhaps. Solar photovoltaics is about the conversion of electromagnetic energy in sunlight by semiconductor materials for our utilization.

--*--

Ch1 references, annotations, supplementary reading

1. The Feynman Lectures on Physics, Conservation of Energy https://www.feynmanlectures.caltech.edu/I_04.html
2. International Astronomical Unition Resolution B3 https://www.iau.org/Iau/Iau/Publications/List-of-Resolutions.aspx
The luminosity of the Sun is given as 3.8 10^{26} J/s. We have converted it to TWh/s - to convey a comparative expression in TWh, since electricity generation and consumption is generally discussed in TWh.
3. International Astronomical Union Resolution B2 https://www.iau.org/Iau/Iau/Publications/List-of-Resolutions.aspx
The perihelion, the closest distance between the Earth in its orbit and the Sun, is 147.1 billion metres. The aphelion, the farthest distance between the Earth in its orbit and the Sun, is 152.1 billion metres. Distances are measured by radar imaging.
4. The Sun's impact on the Earth, World Meteorological Organization https://wmo.int/suns-impact-earth
5. Global Heat Flow: New Estimates Using Digital Maps and GIS Techniques, April 2018, International Journal of Terrestrial Heat Flow and Applications 1(1):6-13, V.M. Hamza and Fabio Pinto Vieira DOI:10.31214/ijthfa.v1i1.6
6. Energy transfer in electrical circuits: a qualitative account, American Journal of Physics, DOI: 10.1119/1.181993, Igal Galili and Elisabetta Goihbarg https://www.ippp.dur.ac.uk/~davis/EnergyTransferElectricCircuits.pdf

7. Energy flow from a battery to other circuit elements: Role of surface charges, DOI: 10.1119/1.3456567, Manoj K. Harbola

Harbola quantitatively describes the energy flow along a long wire carrying DC current, building upon the qualitative explanation in the preceding reference.

8. The Feynman Lectures on Physics, Field Energy and Field Momentum

https://www.feynmanlectures.caltech.edu/II_27.html

Ch2. Rules of the Energy Game

"...and more important, because no one knows what entropy is, whenever you use the term, you will always be at an advantage!!" – John von Neumann

Spontaneous energy conversion processes from one form of energy to another naturally occur. Humans learnt forced energy conversions a long time ago, e.g., burning wood or other materials for thermal energy to keep ourselves warm. Plants convert sunlight into chemical energy. A ball rolled across a surface comes to a halt, after losing kinetic energy through friction as heat. These processes are generally irreversible, finish to start. The fire cannot be converted back to the chemical energy of wood, plants don't return the absorbed electromagnetic energy as received, and heat from friction cannot be inserted back into the ball for a kinetic energy boost. In all of these processes, a common factor is the conversion of some residual thermal energy from the original energy source. Restoring this thermal energy to its original form in the same source seems impossible, at least by reversing the sequence of events. We mentioned the first rule for these processes: the conservation of total energy. If the system boundaries for the aforesaid processes are carefully defined to cover all energy exchanges, the energy ledger accounts shall prove the rule. The total amount of energy remains the same, once the thermal energy is accounted for, but the ability to further use the resulting energy balance has changed. The burnt wood cannot be used in the same way again, the electromagnetic energy absorbed by the plant can now only be used as reduced chemical energy, and the ball will need another push of mechanical energy to resume rolling, eventually exhausting its source. To keep track of how the available energy can be used, we need another rule. The second rule relates to information about the distribution of energies across the energy bearing units of each form in any physical system.

Entropy

For any energy form, its total quantity can be divided among individual energy bearing units or what are called microstates of the energy distribution (microstates can be a bit misleading, as the energy state does not have to be on a micron scale). The number of ways in which the total energy quantity can be distributed across these individual units to specify the distribution is called entropy, denoted by the symbol S (not to be confused with the electromagnetic flux vector).[1] We shall temporarily call it S_{Energy}, to distinguish it from other definitions of entropy, emphasizing its deep-rooted connection to energy. Generally, ceteris paribus, if the number of energy bearing units is increased, so will the possible number of ways of representing the total energy. Entropy is not a measure of the standard deviation in the distribution or dispersion of energies, but uncertainty about the distribution function of the energies. It quantifies what we know about how the energy might be distributed: the less we know, the greater the entropy figure.

Statistical information theory mathematically defines an entropy formula.[2] Entropy is uncertainty about the distribution function of information among the possible information containing units that would completely specify the total information content of the whole system. The information could be about energy or anything else.

The S_{Info} (entropy of information) used in information theory is defined by

$S_{Info} = \sum_i P_i \log_2 (1/P_i) = - \sum_i P_i \log_2(P_i)$

"i": index for the number of possible information containing units

P_i: probability of finding a particular specification of the information unit.

Probabilities are measured in discrete fractions of 0 to 1, on a normalized scale so that the total probabilities sum to 1. In information theory, Boolean bits are used to specify the information, increasing by powers of 2 for the quantity of possible information in each unit, so log to the base 2 is used for $\log_2 (1/P_i)$ the information contained in each unit.

The objective in information theory is to use S_{Info} to find the most parsimonious way to specify the whole system information. In a uniform distribution function, all individual units have an identical probability for containing the same information: each $P_i = 1/N$, where N is the total number of units. For a uniform distribution, S_{Info} reaches a maximum of log(N).

S_{Info} for a uniform distribution = $-\sum\frac{1}{N}\log 1/N = -\frac{1}{N}\sum(\log 1 - \log N) = -\frac{1}{N}\sum(0 - \log N) = -1/N(-N\log N) = \log(N)$.

For a physical energy system, probabilities of energy information can be assigned to individual energy bearing units (the microstates) using normalized real numbers for the probability distribution (when quantum mechanical effects are insignificant), so that:

S_{Energy} = - Entropy Constant • $\sum_i P_i \log_e(P_i)$

"Entropy Constant"[3]: a natural constant for physical information specification in units of energy per degree absolute temperature.

P_i: the probability of finding a particular energy specification in a microstate.

The natural log to base "e" is convenient because in physical systems the rate of change or gradient of a variable tends to be proportional to its quantity (leading to exponential solutions). Lower entropy does not mean there is less to tell, but less ways to tell it.

For an ideal gas in thermal equilibrium, the energy information is uniformly distributed across "N" independent molecules, $P_i = 1/N$, and S_{Energy} is maximised = - Entropy Constant • ln(N).

When quantum mechanical (QM) effects are significant, the physical system entropy S_{QM} has to be formulated in different terms:

$S_{QM} = -\sum_i \eta_i \ln(\eta_i)$, where "η" are normalized eigenvalues of density matrices of quantum mechanical states. S_{Energy} and S_{Info} can be derived from S_{QM} under the limiting conditions when quantum mechanical probability distributions do not have to be used.[4]

The second rule is that for a perfectly isolated system, insulated from any external exchange of energy or matter, a change in S_{Energy} is always greater than or equal to zero: $dS \geq 0$, where dS is short-hand for (dS/dt) dt. Henceforth, we use S for S_{Energy}.

For an isolated system, the unknown information about the distribution function of energies always increases or stays the same when it has reached a maximum. The lower the entropy in an isolated system, the greater the room there is to convert energy to another form, because with every energy conversion step the entropy only increases. For a non-isolated system, entropy can be reduced to make more room but only at the expense of injecting energy directly or indirectly through the interaction with external matter.

This rule sets the available energy calculus for all types of energy conversion processes, revealing hard limits on the maximum efficiency of each process, including that of any form of energy to electricity. It also leads to a fundamental definition of temperature on a scale starting from an absolute temperature of zero degrees K. At 0K, the thermal energy of the system is zero. In practice, we can get close to but never quite reach 0K (at least for the types of energy systems we shall consider). We can never completely get rid of residual thermal energy. The residual thermal energy that cannot be recovered limits the available energy for useful application.

Total Internal Energy and Available Energy

The total internal energy "U" of a physical system is defined by the overarching equation: $dU = T\,dS - P\,dV + \sum_i CP_i\,dN_i$,[5]

"dU": change in total internal energy,

"dS": entropy change

"T dS" ≥ "δQ": the change in thermal energy transfer (a relationship, implicit in the equation)

"dV": change in volume

"T": the temperature

"P": the pressure

"Chemical Potential (CP)": gives the contribution from changes in the number of distinct chemical species indexed by "i". Chemical Potential (CP) has physical units of energy per mole of the chemical species.

dN_i: differential change in the number of moles of species i.

This equation covers the equilibrium states of perfectly isolated (no exchange of matter or energy), closed (no exchange of matter, only energy), or open systems (both matter and energy can be exchanged). It provides a foundation to calculate what energy is available to proceed from one system equilibrium to another.

Depending on the forms of energies being converted, e.g., thermal, mechanical, or chemical, certain state variables can be held constant and the others varied to assess the available energy for useful work. The energy contribution to an increase in entropy, i.e., greater uncertainty about the distribution function of energies, is classified as not useful. Why? Reducing the increased entropy will require additional energy. Entropy debits the available energy. Variants of the total internal energy, fixed by the constraints of the system, can be defined to evaluate the available energy. Minimization of the available energy function maximizes the useful work that can be extracted.

Isothermal energy F, the available energy function for a system held at constant temperature, e.g. by a thermal reservoir and heat exchangers, is defined by $F = U - T S$.

Available energy for useful work is extracted by minimizing F ($dF < 0$), with

$dF = -S\,dT - P\,dV + \sum_i CP_i\,dN_i = 0$, at equilibrium.

The maximum energy extracted for useful work = $-\Delta F$, the total change in the isothermal energy function.

Chemical Potential energy G, available for a system kept at constant pressure and temperature, is defined by $G = H - T S$, where H, the enthalpy = $U + P V$.

Available energy for useful work is extracted by minimizing G ($dG < 0$), with

$dG = -S\,dT + V\,dP + \sum_i CP_i\,dN_i = 0$, at equilibrium.

It is applicable to chemical reactions and phase transitions.

Maximum work that can be extracted for useful work in the form of chemical or electrochemical energy = $-\Delta G$, the total change in the Chemical Potential energy function.

We can use these available energy functions to assess conversions between different forms of energy, testing the fundamental limits and their relation to economic criteria.

Mechanical/Electrical

The mechanical energy of a body is configured by its position, e.g., height of the body, or motion, its mass and velocity. All mechanical energy bearing units of the body have to fit a specific configuration of low entropy set by the macroscopic position or motion of the body as a whole. On the other hand, thermal energy is generated by the energy bearing units randomly moving among a large number of internal configurations, whatever the position of the body. Pure thermal energy, by definition, has the highest entropy. Although the low entropy of mechanical energy enables efficient energy conversion to electrical energy, it is limited by the presence of thermal energy in the system. Consider the available isothermal energy F for a piston expanding air to generate electrical energy ,e.g., in a compressed air energy system (CAES). Since no chemical species is being exchanged, the Chemical Potential term can be ignored. $dF = dU - T\,dS - S\,dT = T\,dS - p\,dV - T\,dS - S\,dT = -p\,dV - S\,dT$. When the piston compresses air, $dF > 0$, conversely when it expands $dF < 0$. If there was no increase in temperature or entropy (dS = 0 and dT = 0), the entire available energy would be converted to mechanical work done for electrical energy, in what would be a reversible process. In reality, irreversible friction between the piston and air causes an increase in both dS and dT for an energy penalty of T dS-irreversible, where dS-irreversible is the irreversible change in entropy from the friction. For a fluid with lower viscosity than air (hence, less friction), the entropy penalty drops for a higher efficiency.

CAES with perfect heat exchangers for an ideal isothermal energy conversion process would achieve a theoretical efficiency of 100%. But this ideal theoretical efficiency is impossible, even for a perfectly reversible process, because the air adiabatically heats up and cools during compression-expansion cycles to set up a parasitic heat engine within the system. The thermal efficiency of the heat engine is limited by $(T_{hot} - T_{cold})/T_{hot}$, where T_{hot} is the temperature of the compressed air, and T_{cold} the temperature of the expanded air. It is an unavoidable thermal loss factor restricting the maximum possible CAES efficiency. Commercial CAES round-trip efficiencies (RTE) are at 70-80%, while prototype systems of isothermal CAES have demonstrated RTE above 95%.[6]

Chemical/Thermal

Chemical energy is stored in highly ordered configurations of electronic states. The simpler the molecular structure storing the chemical energy, other things being equal, the lower the entropy with higher energy conversion potential. In combustion, when the molecular bonds are broken to release thermal energy a large increase of entropy is absorbed by the surroundings. As a result, nearly total chemical to thermal energy conversion is possible, e.g., 98% efficient methane boilers. We shall soon see that for further thermal to mechanical and then electrical energy conversion, i.e., that of a heat engine, there is a much tighter maximum efficiency limit.

Chemical/Electrical

For direct chemical to electrical energy conversion, e.g. in fuel cells, the maximum energy conversion efficiency depends on G, the Chemical Potential energy function. It is given by $\Delta G/\Delta H = 1 - T(\Delta S/\Delta H)$, where ΔS is the change in the whole fuel cell system entropy (excluding the external surroundings). For combined thermal and electric power, the useful energy yield can go higher with waste heat recovery; but the generation of heat loss due to irreversible entropy changes, even for ideal operating conditions, is a limiting factor for the electrical conversion. The maximum theoretical electrical conversion efficiency for a hydrogen fuel cell is 83%, less for alkaline and solid oxide fuel cells. Practical fuel cell efficiencies are much less, mostly to do with chemical kinetics, at present no more than 60% for all three types of fuel cells.

Thermal/Mechanical/Electrical - Ideal Maximum Efficiency

Most electricity worldwide, as of now, is produced by thermal to electrical energy conversion. Chemical or radioactive fuels are used to produce heat that is converted to mechanical and then electric power through heat engines. The maximum possible efficiency of a heat engine is derivable by considering a hot and cold reservoir with a thermal energy transfer from the hot to cold reservoir, that is converted into mechanical work to drive a turbine. This derivation only depends on the first and second rules that we have described for energy conservation and changes in entropy, and a little rudimentary algebra.

The maximum thermal to mechanical efficiency for a single-stage heat engine is given by 1 - (T_{cold}/T_{hot}) multiplied by 100% for a % figure, where T_{hot} is the temperature of the hot reservoir, and T_{cold} the temperature of the cold reservoir. An ideal reversible heat engine process is assumed i.e., the total system change in entropy from the conversion is zero.[8]

This maximum thermal efficiency does not depend on the source of the heat, the substances in either thermal reservoir, the type of mechanical motion device, but only on the respective temperatures. No heat engine system can do better than a reversible one, for a given temperature difference between two reservoirs, no matter what process cycle is designed, whatever combination of isothermal (constant temperature), isochoric (constant volume), isobaric (constant pressure), adiabatic (no heat transfer), or isentropic (constant entropy) processes are used. The derivation of the upper efficiency limit does not involve any reference to or assumption about such processes, other than the reversibility condition.

A reversible heat engine is useful fiction to give us an upper limit of the maximum thermal-to-mechanical energy efficiency between two thermal reservoirs: the heat source and sink. $1 - (T_{cold}/T_{hot})$ is a fixed upper limit for converting the exchange of random internal kinetic energies between two bodies to mechanical energy. The mechanical device design, the heat exchange physical process, and other modalities of the system machinery have been taken out of the equation by the reversibility condition.

It follows that for any heat engine system, we want to increase the inlet temperature of the heated substance driving the engine, and minimise the temperature of the exhaust temperature. These two variables are the headline design objectives. We can only achieve 100% efficiency, by going down to absolute zero temperature for the exhaust - an impossible feat.

A reversible heat engine for maximum efficiency can in theory be implemented by a process cycle of alternating, reversible isothermal and adiabatic actions:[9]

- Reversible isothermal expansion of the fluid in the turbine to T_{hot}

- Reversible adiabatic expansion of the fluid in the turbine to T_{cold}
- Reversible isothermal compression of the fluid in the turbine to T_{cold}
- Reversible adiabatic compression of the fluid in the turbine to T_{hot}; work done in this final step offsets that of the adiabatic expansion

Figure 2: maximum efficiency heat engine cycle

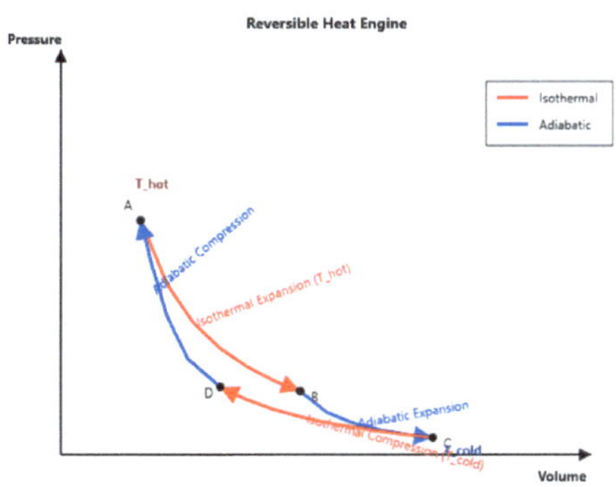

Mechanical work is done in the reversible isothermal steps. A practical attempt would however deliver zero power. Why? An ideal reversible isothermal process is too slow - the infinitesimal, quasi-static changes for reversibility would take an infinite time to do mechanical work so the rate of transfer of energy is zero.

Thermal energy has to be transferred between the reservoirs, by thermal transfer fluids, with a finite thermal transfer time for the process as an upper limit for the thermal transfer rates. Thermal energy is lost in each process cycle step to limit the deliverable power.

An adjusted theoretical maximum conversion efficiency for the maximum heat engine output can be derived for this process cycle (to deliver real-world power), by assuming the adiabatic processes are reversible and equating it to a corresponding maximum rate for the isothermal process. Maximum theoretical efficiency sharply drops to $1 - (T_{cold}/T_{hot})^{1/2}$ by the necessary adjustment. This process cycle is not very practical and only used for theoretical analysis.[10]

Practical heat engines, heated by coal or nuclear radioactive fuel, rely on phase transitions for faster process cycles. Gas engines use a combination of adiabatic and isobaric processes, without phase transitions, in a manner similar to jet engines.

Practical Heat Engines

The global workhorse of thermal to mechanical and subsequently electrical energy generation is the vapour expansion cycle (conversely, heat pumps use a vapour compression cycle). Phase transitions can come close to isothermal reversible processes, e.g., in the melting and freezing of ice at a constant temperature of 273K: the melting of ice absorbs energy in a low entropy to high entropy phase transition for positive enthalpy - the case for solid to liquid/gas, liquid to gas - and then water releases energy during freezing for negative enthalpy. Large thermal energy transfers at useful power rates with relatively low losses become viable. Industrial heat exchanger (HEX) designs play a key role.

Vapour Expansion Steam Engine

The four basic stages of an ideal vapour expansion cycle are:

- Isentropic compression by pump

  Thermal transfer fluid is compressed from low pressure to high pressure

  Enthalpy h1 at pump exit: h4 plus work done by the pump (product of fluid specific volume from condenser, at condenser pressure, and pressure differential)

- Isobaric heating by boiler

  High pressure liquid is heated e.g. to saturated or superheated steam

  Enthalpy h2: of the saturated or superheated fluid at high pressure

- Isentropic expansion through a turbine

  Expansion from high to low pressure drives mechanical work done by the turbine

  Enthalpy h3: of the low-pressure vapour (liquid vapour mix or saturated vapour)

- Isobaric heat exhaust in condenser

  Thermal transfer fluid, e.g. exhaust vapour condensed back to liquid

  Enthalpy h4: of the saturated liquid at low pressure

The maximum theoretical efficiency, thermal to mechanical energy, of this ideal vapour expansion cycle is equal to 1 - (h3 - h4)/(h2 - h1). As for all heat engines, the 1 - (h3 - h4)/(h2 - h1) is less than 1 - (T_{cold}/T_{hot}).

Figure 3: vapour expansion steam engine cycle

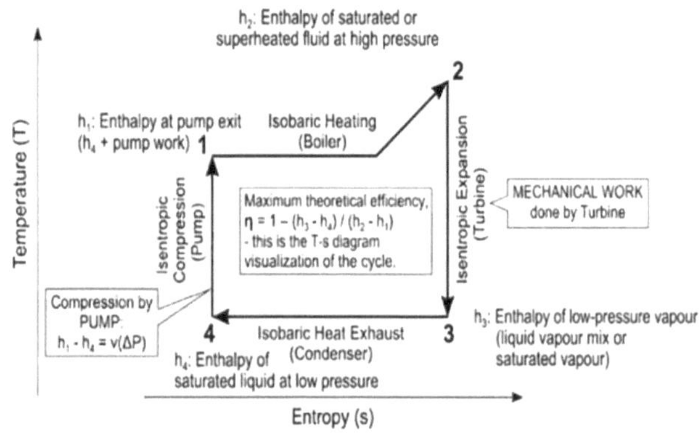

The actual efficiencies of these vapour expansion heat engines are estimated by adjusting the theoretical efficiency for applicable loss factors: the pump efficiency, pressure drops, fluid friction affecting entropy changes, and heat lost to the surroundings from the HEXs. A throttling step is typically applied at the turbine entry in an isenthalpic pressure drop to control the working fluid flow, though it reduces conversion efficiency.

The preferred choice of working fluid that transfers thermal energy through the whole cycle is freshwater. Water absorbs and releases large amounts of energy during phase transitions, and its steam can be superheated to supercritical and ultra supercritical temperatures at high pressure: >800K, 25MPa and 900K, 30MPa respectively. After 647K, the liquid and gas phases of water merge into a single, stable supercritical fluid.

Thermal conversion efficiencies rise from 30-35% for subcritical steam power plants to 40% for supercritical and 45-50% for ultra supercritical. The higher the steam operating temperature, the higher the conversion efficiency. Steam exhaust temperature from the turbine does not change much, about 305-320K, 5-10 kPa, while the steam temperature for turbine entry is raised to drive higher efficiency.

Water determines the vapour expansion power capacity

Thermal conversion efficiency is primarily determined by the working fluid of water, or to be more precise the achievable operating temperature of its steam at the turbine entry. Higher steam temperatures increase the enthalpy h2 to raise the power plant efficiency.

Freshwater is necessary for the steam cycle. Otherwise, the plant machinery would corrode. Around 3,500kg/MWh (subcritical) to 1,600kg/MWh steam flow (ultra-supercritical) is needed to drive the turbine - requiring freshwater withdrawals of approximately 2-4m^3/MWh for the steam cycle. A greater quantity of water is required for the cooling towers. The higher the plant thermal efficiency, the lower the water requirement.[11]

Coal versus nuclear

Thermal-mechanical-electrical power plant efficiency is not determined by the efficiency of the heating source in generating heat from fuel - thermal energy is not the ultimate objective. The critical conversion efficiency step is the transfer of thermal energy from the working fluid - the role of supercritical steam - to the turbine. The heating source only sets the maximum power plant efficiency to the extent it can meet the steam heating requirements. For maximum plant efficiency, the primary concern is whether the source can generate heat to achieve the required steam temperature and then secondly what can achieve this temperature most efficiently at lowest cost. There is an economic trade-off in the cost of fuel to reach higher temperatures.

Nuclear fission fuels cannot achieve the superheating temperatures for ultra-supercritical power plants due to materials safety constraints. Pressure vessel temperature in the nuclear reactor cannot be increased without risking neutron flux damage. Hence, vapour expansion power plants of a higher thermal efficiency are possible with coal than nuclear fuel.[12]

Open Cycle (OCGT) and Combined Cycle Gas (CCGT) Turbines

OCGT[13] and CCGT[14] turbines are designed for an ideal isentropic compression-expansion cycle of thermal to mechanical energy conversion. CCGT uses waste heat recovery to add a vapour expansion power cycle.

The four basic stages of an ideal isentropic-compression cycle are:

- Isentropic compression of air (air instead of water is the thermal energy transfer fluid in this process), to increase its pressure and temperature
- Fuel (typically natural gas) is combusted with the thermal transfer fluid (air), at constant pressure, to yield a high temperature mix of gases
- Isentropic expansion of the hot gas to drive the turbine for mechanical energy
- Isobaric heat rejection of the gases to ambient air, at constant pressure

The maximum theoretical efficiency for an ideal isentropic compression-expansion cycle is given by

1 - (T-ambient/T-compressed) = 1 - (P-inlet/P-outlet)$^{(\gamma-1)/\gamma}$,

where T-ambient: temperature of the ambient air at the compressor inlet

T-compressed: temperature of the air after isentropic compression

P-inlet and P-outlet: inlet and outlet pressures of the compressor

γ: ratio of the specific heat capacity at constant pressure to the specific heat capacity at constant volume of air. This process cycle ideal efficiency is again less than the 1 - (T-cold/T-hot) upper limit for thermal-mechanical conversion.

Actual efficiencies are below the theoretical ideal because of adiabatic heat losses in non-ideal isentropic compression-expansion and pressure drops in the combustion chamber. High-end OCGT efficiencies are around 40%.

Figure 4: OCGT cycle

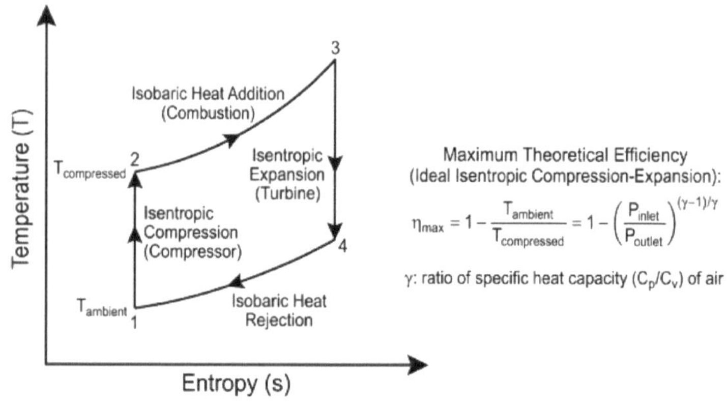

In CCGT, the hot exhaust gas from the first-stage OCGT is used to drive a second-stage vapour expansion cycle on the same principles as described before. Water is boiled for steam for the second-stage cycle. CCGT efficiency is raised to a higher combined efficiency of 60% (first-stage OCGT + second-stage vapour expansion cycle conversion). CCGT offers higher overall efficiency, but OCGT has the advantage of a faster response time because it is quicker to reach thermal equilibrium for the completed process cycle.

We have not discussed the final step for heat engines of converting the mechanical work done by the turbine to electricity. This step shall be discussed in the next chapter as it leads into the detail of electricity networks.

Wrap-up

We have established the basic conceptual foundations for understanding how energy conversion and transfer processes work. The inextricable role of entropy in energy systems has been emphasised. A related engineering concept is that of exergy. It is the converse of entropy: while the total system entropy always increases, exergy is always destroyed (an increase in entropy is a reduction of exergy, and vice versa). Energy efficiency process optimisation aims to minimise exergy destruction.

An understanding of entropy sets out fundamental scientific limits on how various manifestations of energy are convertible from one form to another. It becomes clearer why reducing thermal energy, the highest entropy form of energy, in energy conversion schemes is a crucial priority (except of course if it has to be explicitly used for the purpose of heat).

The natural quality of different types of energy sources is objectively graded in terms of entropy, a yardstick for the useful available energy at our disposal. We are also sensitized to the importance of water in electric power generation, an issue to later expand upon. All matter on the planet is finite apart from occasional space debris, what can be grown renewably with the help of energy, and the Sun is our only source of global energy replenishment.

Entropy in the Earth system inevitably increases from natural or forced energy conversion processes according to the second rule.

Sunlight compensates for the increase in entropy. External energy and or physical matter interactions are necessary to reduce entropy in a given system. There is no significant exchange of matter with outer space, but solar energy is being constantly injected for the growth of life on our planet - making room for further energy conversions and transfers.

--*--

Ch2 references, annotations, and supplementary reading

1. Statistical Physics, Judith McGovern
   https://theory.physics.manchester.ac.uk/~judith/PHYS20352/stat_phys.pdf
   For a more rigorous explanation of the physics of entropy, all the concepts of statistical physics and thermodynamics discussed in this chapter.
2. A mathematical theory of communication, first published in The Bell System Technical Journal, C.E. Shannon
   https://people.math.harvard.edu/~ctm/home/text/others/shannon/entropy/entropy.pdf
3. The Entropy Constant is the Boltzmann constant named after Ludwig Boltzmann by Max Planck, usually given the symbol k in physics textbooks. The most fundamental physical constant of quantum mechanics we mentioned in the first chapter is named after Max Planck: the Planck constant, given the symbol h. Planck named the Boltzmann constant after him in his 1900 paper "Zur Theorie des Gesetzes der Energieverteilung im Normal spectrum" (On the Theory of the Law of Energy Distribution in the Normal Spectrum)
   http://www.ub.edu/hcub/hfq/sites/default/files/planck-energieverteilung.pdf
4. The quantum mechanical entropy was introduced by John von Neumann in 1932 in "Mathematische Grundlagen der Quantenmechanik" (Mathematical Theory of Quantum Mechanics). Lecture notes of Matthew Schwarz on entropy, encompassing the von Neumann entropy and the other formulations.
   https://scholar.harvard.edu/files/schwartz/files/6-entropy.pdf

5. These thermodynamic equations are covered in the Judith McGovern reference. The thermodynamic functions F and G are called the Helmholtz and Gibbs free energies respectively.
6. Advanced Compressed Air Energy Storage Systems: Fundamentals and Applications, published in Engineering, February 2024 DOI:10.1016/j.eng.2023.12.008, Xinjng Zhang et al.
   The review compares the performance of adiabatic, isothermal, liquid, supercritical, and underwater CAES. Isothermal CAES has the most promising RTE.
7. Table listing the molar entropies of various chemical compounds and molecules: https://www2.chem.wisc.edu/deptfiles/genchem/netorial/modules/thermodynamics/table.htm
8. The maximum theoretical heat engine efficiency is called the Carnot efficiency, a result of the Carnot theorem (it is more general than the Carnot engine process cycle).
9. This process cycle is titled the Carnot engine.
10. Efficiency of a Carnot engine at maximum power output. American Journal of Physics, 1975, 43(1), 22-24, Curzon, F. L., & Ahlborn, B. The modified maximum power efficiency for the Carnot Engine is called the Curzon-Ahlborn efficiency.
11. The vapour expansion cycle is known as the Rankine cycle. Description of idealized Rankine cycle process: https://en.wikipedia.org/wiki/Rankine_cycle
12. Rankine cycle efficiency limit for nuclear fission power plants: https://www.nuclear-power.com/nuclear-engineering/thermodynamics/laws-of-thermodynamics/thermal-efficiency/thermal-efficiency-of-rankine-cycle/

Explains the operating temperatures, and hence thermal efficiency, are constrained by the materials of the pressure reactor material.

13. The gas engine process cycle for OCGT and CCGT is named the Brayton cycle. Key OCGT performance parameters (compressor inlet temperature, turbine inlet temperature, and pressure ratio) are explained in:
    Performance analysis of open cycle gas turbines, International Journal of Energy Research, March 2009, Huseyin Kurt et al.
14. Commercial CCGT product descriptions:
    General Electric (GE) technical paper for single-shaft CCGT, in 2000, showing thermal efficiencies reaching 60%:
    https://www.gevernova.com/content/dam/gepower-new/global/en_US/downloads/gas-new-site/resources/reference/ger-3767c-single-shaft-combined-cycle-power-generation-system.pdf
    Recent GE product line:
    64% efficiency:
    https://www.gevernova.com/gas-power/products/gas-turbines/9ha
    Siemens turbine that can be used in OCGT or CCGT for 40% and 60% efficiency respectively:
    https://www.siemens-energy.com/global/en/home/products-services/product/sgt-800.html#/

Ch3. AC Power Grid

"If you want to find the secrets of the universe, think of energy, frequency, and vibration" – Nikolai Tesla

We spent some time on the limits of converting thermal energy, arising from random motion within physical bodies, into orderly mechanical energy that could be put to useful work. Entropy, via the second rule, dictates a strict upper limit on the efficiency of converting the highest entropy form of energy to a lower entropy form. Even if the heating fuel was vastly abundant and cheap, the working fluid of water to maintain thermal energy transfer is a critical resource with urgent competing needs for humans.

Converting low-entropy mechanical energy to electrical energy, for the next stop, ought to be less taxing - think of entropy as a built-in tax on the available energy. The process of using a piston with air for compression and expansion was discussed and found rather limiting because of friction, as well as the adiabatic heating and cooling giving rise to a parasitic heat engine with thermal energy losses. The unavoidable thermal energy loss from the adiabatic temperature variation scales as $(P_{initial}/P_{final})^{(\gamma-1)/\gamma}$, where $P_{initial}$ and P_{final} are the initial and final pressures after compression or expansion, and γ is the ratio of the specific heat capacities at constant pressure to constant volume of the fluid medium. Pressure variation exacerbates frictional heat loss.

Electromagnetic Generator

The interplay of electric (E) and magnetic (B) fields in electromagnetic energy, i.e., the two-way process of electromagnetic induction, provides an alternative solution. Relative mechanical motion set up between two metallic conductors of stationary and rotating parts, called the stator and rotor respectively, converts mechanical to electrical energy avoiding the adiabatic heating and cooling of piston movement. Setting up the mechanical work done as rotational kinetic energy incurs less friction than the linear motion between two surfaces, because of reduced contact area, lower friction coefficients, and smoother motion.

The stator consists of a metal core (a ferromagnet) for magnetization, metallic conductor windings in a geometry to precisely shape the induced E-B fields, with insulation for protection, and cooling structures (e.g. fins, dielectric jackets). Thin sheets of laminated silicon steel are housed in a cylindrical steel casing for the magnetic core, with slots for the windings, teeth and yoke construction to steer and complete the magnetic loops. Stacking of laminated steel sheets in the core reduces eddy currents.

The rotor is a cylindrical shaft, either a permanent magnet or surrounded by magnetic coils, made out of high strength steel alloy to withstand fast angular velocities without vibrating, with embedded or protruding poles to set the frequency of the induced

E-B fields in the stator. It is placed inside the stator with a narrow air gap for maximum inductive flux without the risk of contact. With this arrangement, the mechanical to electrical conversion efficiency is 95-99%. Thermal to electrical losses of heat engines are dominated by the thermal to mechanical intermediate step (chemical or nuclear to thermal being the initial step).

Equations summarizing the whole process:[1]

P_m (mechanical power) = thermal efficiency • thermal power

While powering up the mechanical generator:

$P_m = (I \cdot \alpha \cdot \omega)$ + energy losses from friction, magnetization, coil resistance, total losses of typically 1-5%

"I": moment of inertia

"α": angular acceleration

"ω": angular velocity = 2πf, f for frequency

"I • α": the mechanical torque, torque-mech

During steady-state motion, ignoring the relatively small 1-5% electromagnetic generator losses from turbine friction, the magnetic core and windings:

Thermal-efficiency • thermal power = $P_m = \text{torque}_{mech} \cdot 2\pi f$

Mechanical torque "torque_{mech}" generates an electrical torque "torque_{elec}" $= k \cdot B_r \cdot B_s \cdot \sin(\delta)$

"k": a design constant that factors in the number of poles, coil areas, number of turns, air gap between rotor and stator, geometry of the machine

B_r: magnetic field of the rotor

B_s: magnetic field of the stator

"δ": the torque angle between the stator and rotor magnetic fields

In an open-circuit condition, the heat engine is throttled after delivering a kinetic energy impetus. A trickle of thermal power sustains the mechanical motion against friction and magnetic core losses.

Connecting the generator in an electrical circuit to a load, delivers:

$P_m = torque_{mech} \cdot \omega = P_e$, the electrical power

$P_e = V \cdot I \cdot \cos \phi$, where ϕ is the phase angle between the voltage and current aligned to the load; for a purely resistive load $\cos \phi = 1$.

Voltage (V) and current (I) have time-dependent sinusoidal AC waveforms from the oscillating E and B fields in the generator, with the temporal variation automatically synchronized to the rotational motion of the turbine. This electromagnetic generator conversion process with a high efficiency of 95-99% is naturally suited to delivering AC power.

Adapting the generator for DC power requires the complication of a commutator with additional cost, power loss, and stability issues. A commutator consists of copper conductors cylindrically mounted on the rotor shaft, connecting the ends of the windings with stationary carbon or graphite brushes to rectify the current flow to and from the external circuit. Electrical connections are reversed in sync with the turbine rotation for DC current. The pulsating DC current from the turbine rotation is smoothed out across multiple connections. Brushes get worn out with time. AC electromagnetic generators have simpler designs, higher efficiency, and longer stability than the DC adaptation.

The historical reliance on heat engines for electricity, with generators integrally suited to AC power conversion, is a major reason why primarily AC electricity networks have been built out worldwide.

AC Advantages and Challenges

We begin with a broad-brush list of the functional requirements for a universal energy sharing system, to assess how electricity in particular AC power would serve the purpose.

Compatibility with useful energy sources:

Energy source utility in scientific terms can be graded according to entropy. Scientific grading has to be tempered by the application requirement; e.g., though thermal energy has the highest entropy, it might be a necessity for heating.

The entropy rule shows that low-entropy sources are more easily convertible to higher entropy forms (lower entropy tax of thermal losses). Therefore, an energy sharing system that is suited to integrating, transmitting, and distributing low-entropy forms of energy is desirable.

Electricity is a low-entropy form of energy. Thermal energy is high-entropy, but convertible to electrical energy. We have seen that efficient thermal to electrical conversion is undertaken by AC power generation.

Compatibility with user energy loads:

The energy delivery is readily adaptable to a variety of consumer appliances and use-cases. Electrical load specifications can be standardized for diverse applications, and the power delivery characteristics precisely controlled to meet these well-defined specifications. Consumer networks and commercial appliances evolved for AC power, so did the associated safety standards and certifications, because it suited the electrical transmission grid.

Speed of power transmission:

The pace of power delivery varies for different energy forms. Electromagnetic transfer is the fastest possible - close to the speed of light. Transmission speed of energy transfer does not slow down over long distances. The speed may vary according to the cable design, how it is constructed, but it does not depend on the length of the cable.

Capacity and scalability:

The medium for power transfer must be capable of high-power capacities and scalable for long distance networking. Direct gas pipelines are also capable of high-power capacities but are more capital intensive to scale. There is greater flexibility in how to geographically plan and build out electrical grid lines.

Transmission and distribution losses:

Minimizing transmission and distribution losses for efficient power transfer is of paramount importance - it affects the capacity, scalability, and economy of the energy sharing network. The interplay between voltage and current (associated electric and magnetic fields) - an ability to manipulate one or the other for the same power product - enables high-voltage power transfer thereby reducing the electrical resistance losses of thermal energy. We shall later see how modern electronics, including the interfaces of solar photovoltaic systems, can help maintain the required voltages for power transfers across the electrical grid. Voltage drop with distance is a greater issue for AC power transfer - unlike DC, there is reactance in addition to resistance.

Control of stability, reliability, and safety:

Control is a strong suit of electricity. A variety of control mechanisms are possible for AC and DC power transmission. In fact, the whole internet is based upon precise control of electric power transmission coded in specific information protocols. In ICT infrastructure, the objective is maximum throughput of data with minimum power consumption. For electric power transmission, the goal is maximum power throughput.

The power supply for our energy sharing network must be stable, reliable, and controllable to the maximum extent possible at all levels including transmission, distribution, and end-use. Redundancy and emergency response mechanisms are essential so that energy demand is met by sufficient supply at all times without any interruptions or system-wide breakdowns.

Consumer demands are constantly changing: the amount of energy required as well as its purposes, and the conditions for operating the energy sharing network. Flexibility is therefore highly valuable. Resilience to extreme weather events and fast recoveries from accidents are important planning factors. Safety considerations include environmental impact.

Economy and sustainability:

All of the above must be economical not only in terms of short-term financing calculations but long-term sustainability. The long-term sustainability depends on efficient use of finite materials resources, compatibility with materials recyclability, and natural constraints.

Single and Three-Phase Compatibility for Loads

With this functional overview in mind, we return to the AC power output from the electromagnetic generator of the heat engine - to see how it meets the simple resistive load of an old, incandescent lightbulb (ignoring capacitance and inductance, for the time being). The power output of the generator, its voltage and current, is synchronised in frequency to the rotational motion of the turbine, so these types of power sources are called synchronous generators. The voltage and current of the power output are given by

$V(t) = V_{max} \cdot \sin 2\pi f \cdot t$ and $I(t) = I_{max} \cdot \sin 2\pi f \cdot t$ ($\phi = 0$, $\cos \phi = 1$).

Electrical power supply "P" (active electrical power is denoted P, henceforth), V(t), and I(t), are oscillating with frequency f, but we don't notice it in the lightbulb. The reason is the time taken by the filament load, T_{load}, to integrate the P input into visible light for our eyes: $T_{load} >> 1/2f$. For 50Hz, 1/2f = 10ms (ten milliseconds), while the bulb response time T-load = 100-200ms.

AC power P = "W" watts rating of a bulb, with resistance "R", the requisite power delivery for the time taken to light the filament.

$V_{max} = (2W \cdot R)^{1/2}$, $I_{max} = (2W/R)^{1/2}$

Root-mean-squared voltages and currents:

$V_{rms} = V_{max}/2^{1/2}$, $I_{rms} = I_{max}/2^{1/2}$

$V_{rms} \cdot I_{rms} = W$ (the lightbulb power capacity rating)

AC power P flowing in a single conductor line has the time dependence:

$P(t) = V(t) \cdot I(t) = V_{rms} \cdot I_{rms} \cdot (1 - \cos(4\pi f \cdot t))$.

When 1/2f << T-load, as in old incandescent lightbulbs, the oscillatory P, V, I do not matter - the oscillations average out to meet the load rating within its response time. P(t) integrated over T_{load} seconds = "W" watts $\cdot$ T_{load} seconds, so that the required energy gets through in the required time without anyone noticing the difference for shorter periods.

Loads with greater T_{load} sensitivity require minimization of the oscillatory power effect. AC power is equally split among three conductor lines inside a power cable (line power = P/3), and simultaneously delivered from the generator source to the load.

V(t), I(t) of each line are fixed at a constant 120 degrees phase difference in time to the other. Constant power is supplied to the bulb load W at all times: $W = P(t) = 3 \cdot V_{rms} \cdot I_{rms} \cdot \cos \phi$. A purely resistive load has $\phi = 0$.

Three-phase systems are balanced for constant power (P) delivery in - the voltage magnitude between each line and the current magnitude in each line are identical. Total power P is delivered evenly, with P/3 on each conductor. When the load is potentially unbalanced, e.g., a mix of high-power three-phase loads (industrial motors) and many smaller, single-phase loads (residential electronics), a neutral line is added for a safe return path of any unbalanced currents.

Theoretically, there isn't a problem in three-phase AC delivery for constant power output to meet shorter load response times, but there are practical implications. The more conductors involved in the electrical circuit, the greater the number of possible faults in and between conductor lines. When single-phase AC suffices for the load response time, other things being equal, it's a lower maintenance risk.

Reactive Power (Q), Power Factor (PF), Voltage Support Q(V)

For AC, purely resistive conductors and loads don't really exist (though used for idealized approximations like the incandescent lightbulb). Impedance "Z" applies to AC, where $Z = \{R^2 + (2\pi f \cdot L - 1/2\pi f \cdot C)^2\}^{1/2}$, "R" is the Ohmic resistance, "L" the inductance, and "C" capacitance. Z only reduces to R for no electrical frequency i.e., DC. The impedance component $(2\pi f \cdot L - 1/2\pi f \cdot C)$ is called reactance: inductive (L) and capacitive (C) corresponding to the B(t) and E(t) fields associated with current and voltage.

AC power is comprehensively defined by total Apparent Power "AP" $= (P^2 + Q^2)^{1/2}$ where Q for reactive power is a tax on the AC power delivery, counted in units of VAr, P the active power in watts that can be used by the circuit load.

Power Factor (PF) is defined by P/AP. It provides a measure of the necessary reactive power to sustain AC power delivery. The phase angle difference "θ" is related to PF by $\cos\theta = P/AP$, equal to 1 when V and I are in phase to transfer maximum active power, and 0 when V and I are 90 degrees out of phase to provide reactive power only. In general, $V(t) = V_{max} \cdot \sin(2\pi f \cdot t - \theta)$ and $I(t) = I_{max} \cdot \sin(2\pi f \cdot t)$, where θ is the phase angle difference between V(t) and I(t).

Total apparent power AP, active power P, and reactive power Q are related by

$AP = V_{rms} \cdot I_{rms} = V^2_{rms}/Z$, where Z is the absolute impedance magnitude.

$P = V_{rms} \cdot I_{rms} \cdot \cos\theta$, $Q = V_{rms} \cdot I_{rms} \cdot \sin\theta$

Inductance causes voltage drops in AC power transmission lines with distance, reducing the PF and hence leading to loss of active power (greater current is drawn for the voltage drop and electrical resistance losses scale as the square of the current). To compensate for this effect, the provision of Q is necessary.

Reactive power substations are installed across the AC electric grid network for local voltage support, denoted Q(V): an injection/absorption of Q for voltage stabilization within a target voltage range. PF is kept less than 1 for voltage support, typically ~ 0.9-0.95. Lacking adequate Q(V) provision from the substations and or grid-connected generators, the whole grid system can collapse (and sometimes does, like recently in Spain and Portugal). When X/R is high, where X is the line reactance, the transmission line (e.g., in rural distribution grids) is weaker with a lower $\cos \theta = 1/(1 + (X/R)^2)^{1/2}$. There is an increase in power loss from a voltage drop or rise for inductive and capacitive reactance respectively. The higher the line X/R, the greater the need for Q(V) to support the AC power delivery. For X/R of 0.5, PF falls below 0.9. Inductive and capacitive effects are stronger in overhead and underground lines respectively. DC is preferred over AC for undersea or underground cables longer than ~50km, due to higher capacitance between the conductor voltage and cable grounding shield.

Voltage magnitude as a function of time and the waveform frequency are coupled together by $V(t) = V_{max} \cdot \sin(2\pi f \cdot t)$, with a deviation in one parameter affecting the other. For stable synchronous power flow in an AC circuit, these fundamental electrical parameters of voltage and frequency, affected by inductance and capacitance, have to be carefully managed together. Q(V) is used for voltage support but if the frequency collapses, so will the voltage. Active power P is used to support frequency but conversely, if the voltage goes out of control, so will the frequency.

Skin Effect

Another frequency-dependent AC effect that increases the electrical resistance to power flow is the "skin effect". AC current density is concentrated near the surface of the conductor, with a skin depth given by

Skin depth = (resistivity/(π • magnetic permeability • frequency))$^{1/2}$, defined by an exponential decay in current density for that depth - i.e., by a factor of 1/e (~37%),

Skin depths for copper and aluminium at 50Hz are around 9mm and 12mm respectively – therefore at this frequency, low-voltage (LV) cables with comparable diameters are not significantly affected. But for medium-voltage (MV) and high-voltage (HV) cables, with much larger conductor cross-sections an AC cable needs more copper or aluminium material to achieve the same resistance as DC. Alternatively, the skin effect has to be mitigated by special cable designs using stranded wires or hollow cores. Either way, the result is AC cables becoming more expensive than DC for the same voltage rating. AC cable insulation needs to be rated for a higher V_{max}, with $V_{max} = 2^{1/2} \cdot V_{rms}$ (the DC equivalent voltage). DC cables for the same voltage rating are cheaper and lighter in weight. A 500kV AC cable using an aluminium conductor is about 20% more expensive per unit length than an aluminium DC cable of the same voltage rating.

The skin effect does not necessarily mean AC power transmission for the same distance is more expensive than DC as the cable is not the only hardware cost for consideration - voltage conversion substations are a substantial cost difference between DC and AC. DC is cheaper for the same length of power transmission lines when no voltage conversion is required for the load or after a distance of ~600km for overhead lines even if HV/MV conversion is required (generally, voltage conversion is necessary).[2]

HV/MV/LV grid conversions

Reducing transmission and distribution losses is a high priority for energy sharing grid networks. For electricity, the transmission losses are inversely proportional to the square of transmission voltage so for long distances HV is used (ultra-high voltage cables have reached 1100kV). Distribution grids need a voltage step-down for safety to MV and then LV for the end-consumer's habitat. Time-varying magnetic fields in AC power transfer provide a convenient method to step voltage up and down. Since DC involves time-independent steady-state E-B fields, this electromagnetic induction method is not possible.

HV/MV, MV/LV transformers use the same basic physical principles as in an electromagnetic generator but without the need of any moving parts (the electromagnetic power flow already has time varying E-B fields). A wide range of voltages are covered from 1100kV to 110V by the same method. Primary and secondary metallic coil windings are wrapped around a magnetic core, e.g., stacked sheets of laminated silicon steel, for voltage conversion according to the equation:

(Primary voltage/Number of primary conductor turns) = (Secondary voltage/Number of secondary conductor turns).

Power losses in the process are only one to five percent. Electromagnetic induction without any electrical contact between the primary and secondary sides avoids electrical resistance losses (aka Ohmic resistance losses, e.g., in a resistor voltage divider) other than that for the windings magnetization current, and there is no friction of any moving parts. Windings use thick copper to minimize resistance losses in the primary and secondary. We shall discuss more detail about the practical features and selection criteria for transformers and switchgear in a later chapter.

Electromagnetic power flow travels in the proximity of the speed of light ~10^8m/s along or within transmission cables, depending on whether the conductor is insulated or not - either air or dielectric insulation is the electromagnetic propagation medium. Within a transformer, the E-B fields from the primary to the secondary flow through the silicon steel core with a speed that is two orders of magnitude slower ~10^6m/s. The power transfer speed is determined by magnetic permeability (electromagnetic propagation speeds are set by the product of dielectric and magnetic permeabilities, in this case dominated by magnetic permeability). These power transfer speeds far exceed what is possible with other forms of energy.

DC semiconductor technology has progressed to step up/down voltages for high-speed energy transfer, the innovation of solid-state thyristors being the main breakthrough for HV/MV conversion. This technology is more expensive as yet though costs are falling with industrial scale-up. HV/MV substations for DC cost more than the AC versions using transformers. So even though DC cables are more materials efficient and cheaper than AC, the cost of DC HV power transmission is higher for certain distances. Presently, at an overhead transmission distance of ~600km, the DC cable economy and lack of reactive power compensation for DC outweighs the greater cost of solid-state HV/MV substations. The financial crossover distance between long-distance AC and DC is being reduced by falling costs of solid-state HV/MV conversion technology.

HV, MV, and LV network layouts

Electricity transmission and distribution AC grid networks evolved along the lines of a hub and spoke model with large, centralized power plants for generation hubs. Larger centralized power plants, mainly thermal-electrical and mechanical-electrical hydropower, could deliver much lower electricity generation costs than those possible for smaller sized systems - this situation no longer holds true thanks to new distributed energy resources (DER).

HV transmission networks typically use what is known as meshed circuit topology: multiple interconnected paths between the power plants and the consumption centres. Multiple connected paths provide redundancy for unexpected outages of a single power plant. Minimum N+1 redundancy is incorporated, so that the grid is resilient to the disconnection of any single power plant, N stands for the total number of power plant connections and 1 shows any single power plant disconnection is redundant (the acid test is disconnecting the largest power plant or electrical supply line).

A mesh of interconnected paths requires more complex planning, in particular locating faults, analysing these faults, and protecting the whole circuit against the propagation of faults along unintended paths. Hard and soft "open points" have to be carefully planned, an open point being a circuit-breaker or switch to isolate electrical paths, hard versus soft referring to whether the disconnection is a complete physical break or controlled power flow, analogous to an open/shut or sluice gate, respectively.

For MV distribution networks, a ring design with limited meshing is more common. A ring design means a closed, circular electrical loop. The purpose is to make it easier to locate and isolate faults.

LV networks adopt more of a tree and branch layout, called a radial topology, where a single power source feeds multiple loads through a tree feeder. The tree and branch setup is convenient for planning protective circuit-breakers, switches, and fuses. As electricity networks expand with more diverse loads, the introduction of DER for bidirectional power flows, meshing is increasing in MV and LV networks with greater challenges for reliability in particular stable voltage and frequency regulation. Voltage regulation has to be controlled locally, with DER providing new local capabilities for it, while frequency is a global control parameter.

Grid Frequency

Synchronous power flow through the AC grid is essentially an electromagnetic wave of energy at a single frequency of its time varying E-B fields. Synchronicity, as in a single frequency across thousands of miles of AC transmission and distribution networks, is a critical unwavering necessity. Even though the power in the electromagnetic wave can be precisely split up, with efficient step changes in voltages for its transmission and distribution, a single frequency has to be maintained across the entire grid network. Otherwise, the steady, sinusoidal time variations of its E-B fields become unstable. A number of electromagnetic waves with differing frequencies will emerge in a complex pattern of unpredictable, uncontrollable power flows.

The frequency dependent reactance term, $X_L = 2\pi f \cdot L$ or $X_C = 1/2\pi f \cdot C$, dominates the impedance Z of AC transmission lines at 50/60Hz. For wide frequency variations, impedance mismatches would develop across the entire grid network disrupting power flow. Power flow can get reflected back across parts of the circuit, or form a standing wave that destroys the circuit. Reactive power compensation on the network, tuned for a particular frequency, would fail and with it the voltage regulation architecture for a system-wide breakdown.

For the case of only two markedly different frequencies f_1 and f_2, $P(t) = V_1 \cdot V_2 \cdot \cos 2\pi(f_1 - f_2)$, an electromagnetic wave with a beat frequency of $(f_1 - f_2)$ can propagate at close to the speed of light, across a network that is not designed for it (note the line voltages V_1, V_2 will also differ, for different frequencies). Such wave propagations with unwanted frequencies are called frequency instabilities. Resonant harmonics ruin the single frequency waveform power quality. Electromagnetic pulses, interference and destruction of energy ensue, crashing the grid with enormous damage to its equipment and loads. In an AC grid, power flow and its frequency cannot be decoupled across various network segments without an intermediate DC conversion (AC/DC/AC) - except by the use of technology like cycloconverters (can synthesize modified frequencies from the power input for the power output in AC/AC power conversion).

Some frequency instabilities arising from generator faults, congestion and other grid dynamics, can propagate as electromechanical waves with lower frequencies 0.1-2Hz and slower speeds on the order of 100-1000km/s. A frequency instability might resonate with a mechanical torsional mode of a power plant turbine resulting in sub-synchronous oscillations and damage. Synchronous generators need a narrow stable frequency bandwidth around a single specified frequency for safe and reliable operation.

Global frequency control is the main technical challenge, or handicap if you will, that distinguishes AC from DC power grids. There are no frequency dependent effects in DC, including reactive power compensation and grid synchronicity requirements at a single frequency. Frequency is the lynchpin of the AC grid.

Synchronous Generator Operation

Synchronous generators (SG) have to operate at the specified grid voltage and frequency, with narrow bandwidth tolerances for each, and responsive capabilities to stabilize the grid. In worst-case scenarios, to prevent damage from an out-of-control grid, the SG shuts itself off. The SG is designed and constructed for a given range of apparent power (AP) capacity, operating voltage, and frequency.

For a three-phase SG, $AP = 3 \cdot V_{phase} \cdot I_{phase}$, where V_{phase} and I_{phase} are the voltages and currents in each phase respectively; phase current being the stator armature current.

$V_{phase} = 4.44 \cdot f \cdot N \cdot k \cdot \varphi$.

4.44 is an arithmetic simplification of root-mean-square values = $2^{1/2} \cdot \pi$

k: a winding correction factor for the geometry of copper coils across the magnetic poles

f: frequency

N: the stator winding turns per phase

Φ: the magnetic flux per pole

$I_{phase} = J \cdot A_{stator}$.

J: current density of the copper stator windings

A_{stator}: total cross-sectional area of the windings for that phase

The product $AP = 13.32 \cdot f \cdot N \cdot k \cdot \varphi \cdot J \cdot A_{stator}$ is the apparent power generated in units of Volts-Ampere (VA).

The upper limits of the right-hand side parameters, and hence the maximum AP capacity (AP_{max}) are fixed by its design and construction.[3]

Automatic Voltage Regulation

SG voltage is set for the grid voltage target range and controlled independently of the thermal power supply and rotational motion of the turbine. Automatic voltage regulation (AVR) is enabled to sense the grid voltage deviations and adjust voltage to compensate for it.

If the rotor has field windings, the AVR is implemented through an external DC excitation current to the rotor windings for varying the magnetic flux density of the coils. For a permanent magnet rotor, the AVR uses power electronics of AC/DC/AC conversion to independently set the voltage, decoupled from the grid and the rotor motion, by controlling the stator currents and hence its magnetic flux (rather than that of the rotor).

Steady rotor motion, at a synchronous speed, ensures a smooth oscillating magnetic field. The strength of the magnetic field, and hence voltage, is determined and controlled by the independent excitation current magnetizing the rotor windings. Reactive power absorption or injection is adjusted by the excitation current from an external power supply, to set and control voltage.

For a permanent magnet rotor, the magnetic flux density is fixed and cannot be changed. This voltage control is 10x faster than rotor field AVR - a typical response time of 100ms versus 1s, because power electronics works much faster than electromechanical action. The inductance of the rotor windings is also a rate limiting factor. Inductance increases with the size of the windings. Power electronics is faster than rotor AVR, but less flexible because the fixed rotor magnetic flux density narrows the voltage control range. Most thermal power plants use rotor windings, rather than permanent magnet rotors, for greater operational flexibility.

Generators with rotor windings are heavier than those with permanent magnets. The greater mass provides more mechanical inertia. This mechanical inertia proves useful to slow down frequency disturbances. So does the inductance of the rotor windings as an electrical damping factor.

Inertia and Frequency Regulation

The mechanical inertia (MI) of the SG, called the "inertia constant" for the machine, is defined by MI = $(\frac{1}{2} \cdot M \cdot 4\pi^2 \cdot f^2)/AP_{max}$.

M: moment of inertia of the generator fixed by the distribution of masses in its physical shape

AP_{max}: maximum apparent power capacity of its design and construction

f: the frequency set by the angular velocity of the rotor

The physical units of MI are seconds. M and AP_{max} are fixed. MI is directly proportional to the square of the turbine angular velocity, and hence f^2. MI measures the theoretical time the rotational kinetic energy of the SG could supply AP_{max} on its own if the thermal power supply (known as the prime mover) was suddenly cut off completely.

Dynamically (from F= ma), $M \cdot d^2\theta/dt^2 = torque_{mech} - torque_{elec}$.

θ: angular position of the rotor

$Torque_{mech}$: mechanical torque on the rotor

$Torque_{elec}$: electrical torque on the rotor exerted by the electrical circuit

$\theta = 2\pi \int f_{mech}\, dt$, where f_{mech} is the SG mechanical frequency, $2\pi f_{mech}$ the angular velocity in radians per second.

Mechanical power P_m = Torque-mech $\cdot\, 2\pi f_{mech}$

Electrical power P_e= Torque-elec $\cdot\, 2\pi f_{elec}$

We define f_{sync}, synchronous frequency, as the mechanical frequency of the SG that sets the target electrical frequency of the circuit, so that f_{sync} = synchronous speed in RPM • (number of poles)/120, the 120 factor an arithmetic adjustment for RPM to Hz.

$M \cdot d^2\theta/dt^2 = torque_{mech} - torque_{elec}$ can be re-written by substituting:

M in terms of MI and AP_{max}

Synchronous $f_{mech} = \omega_{m,sync}/2\pi$ for f

Electrical angle δ = (number of poles)/2 • θ

We obtain $(4 \cdot MI/\omega_{m,sync})(AP_{max}/poles)\, d^2\delta/dt^2 = P_m - P_e$.

A damping coefficient "D" for friction, and electrical losses in the windings is introduced for the full equation - called the "swing equation" for a SG (thinking of the electrical angle like a swinging pendulum with damping):[4]

$(4 \cdot MI/\omega_{m,sync})(AP_{max}/poles)\, d^2\delta/dt^2 = P_m - P_e - (D/\omega_{m,sync})\, d\delta/dt$.

The electrical angular velocity $\omega_{e,sync} = (poles/2) \cdot \omega_{m,sync}$.

The swing equation in terms of the electrical angular velocity becomes

$(2 \cdot MI \cdot AP_{max} \cdot d^2\delta/dt^2)/(\omega_{e,sync}) = P_m - P_e - (D/\omega_{e,sync})\, d\delta/dt$.

We have absorbed 2/poles in an adjusted damping coefficient D on the right-hand side.

To maintain a synchronous speed for a given electrical power load, we obtain $P_m = P_e + (D/\omega_{e,sync})$.

In practice, the electrical frequency of the grid is fluctuating around the target frequency of 50Hz/60Hz, with loads switching on and off to change P_e. The SG control adjusts P_m to restore the synchronous frequency ω_{sync} – keeping it within a narrow bandwidth. Frequency regulation is provided by adjusting the active power supply from P_m to the grid. A system-wide control process of automatically adjusting active power supply across the SGs on the grid for frequency regulation is called Automatic Generation Control (AGC).

The inertia MI is a measure of the immediate SG response to slow down frequency deviations. The greater the rotational kinetic energy of the SG (increasing with a higher moment of inertia M), the greater its inertial effect in slowing down the frequency deviation.

Although the higher MI (from larger M), slows down grid frequency fluctuations, it also slows the SG ability to ramp P_m up/down to balance active power and quickly regulate frequency. Thermal inertia, i.e., the completion time of the steam or gas process cycle, adds to the P_m response time. Thermal inertia does not contribute to stabilization of transient frequency deviations but it does slow down frequency regulation response times.

Frequency regulation is classified by response times: sub-second fast frequency response (FFR), primary frequency response (PFR) on the order of a few seconds, secondary frequency response of AGC on the order of tens of seconds to minutes, and then frequency restoration response (FRR) on a longer timespan for persistent frequency deviations. Greater inertia in a generator restricts it from FFR, but it can participate in longer restoration responses.

Typical "inertia constant (MI)" values, measured in seconds, for different types of SG machines:

Strictly speaking, the total SG capacity is rated in MVA but commercially MW is often quoted.

OCGT 1-300MVA: 1-5s.

It has lighter rotor mass, and a faster gas heat engine than a steam cycle for the prime mover. The time for full frequency response: 2-10s.

E.g., OCGT power plants in UK: 50MW-140MW.[5]

CCGT 300-800MVA: 4-7s.

The added mass and vapour expansion steam cycle of CCGT increases its mechanical and thermal inertia respectively, limiting its capability for fast PFR. It plays a larger role in SFR.

Time for full frequency regulation response: 15-30s.

E.g., Mitsubishi 500MW, Takasago, Japan.[6]

Coal 300-2,200MVA: 4-10s.

Time for full frequency regulation response: 5-15s for advanced coal power plants, more commonly 15-30s.

E.g., Drax 600MW, UK, or the group of 4,400 MW in Neurath, Germany.[7]

Nuclear 300-1,600MVA: 6-10s.

With nuclear power plants, there is the additional time-consuming complication of stabilizing radioactive byproducts. Hence, limited to participation in FRR.

Time for frequency regulation engagement is on the order of several minutes.

E.g., Flamanville 3 1,600MW, France.[8]

Hydro 50-1000xMVA: 3-6s.

These plants may use very heavy rotors but with relatively low inertia normalized to total apparent power capacity. Conventional hydro has frequency response times of 1-10s, but innovative variable speed turbine designs, rotor decoupling from water flow, and pumped hydro capability can provide FFR below a second.

E.g., Grand'Maison 1,800MW, France.[9]

Grid operators express total synchronous inertia on the grid in units of MW • seconds to factor in the MW capacity of the available machines.

A few seconds can prove to be more than enough for grid stability to unravel. SGs are vulnerable to large fast frequency deviations ,e.g., from HV cable discharge (a tree falling on the cable, extreme weather, or a latent defect), several large power plants unexpectedly tripping, large industrial loads suddenly aggravating power imbalance, or a combination thereof. A fast fault current can suddenly crank the electrical angle acceleration, $d^2\delta/dt^2$, inducing torsional stress in the rotor shaft, leading to mechanical fatigue and cracking. Slow P-m response makes it worse. A large fault current spikes P_e causing excessive heating in the windings and damaging the insulation. Pole slipping beyond 90 degrees with a loss of synchronicity trips SGs for a cascading grid failure. SGs can produce fault current currents of much greater magnitude: 5-20x their rated current capacity (for contrast, solar power plant faults are limited to 1-2x the rated current).

The 1977 New York blackout from a lightning strike on a HV cable, produced a fault current spike tripping 2,100MW, cascading into over 6,000MW with the grid frequency going out control. New York was blacked out again in 2003, from voltage sags, line failures, leading to frequency collapse. [10,11]

Sub-synchronous oscillation (SSO) is a latent frequency-dependent risk. Reactive power compensation circuits on the grid can oscillate at frequencies below the synchronous grid frequency. The SSO frequency can couple with 10-40Hz torsional modes of the rotor shaft, forcing higher amplitude oscillations, eddy currents, overheating, and severe mechanical damage. SSO became a problem for the 600MW Shangdu coal power plant, inner Mongolia, China, in 2011.[12]

Inter-area oscillations (IAO) that typically occur in a frequency range of 0.1-2Hz are a bigger problem.[13] IAO are electromechanical waves propagating at 100-1000km/s, slower than fast fault currents close to the speed of light, but can also cause severe grid malfunctioning including a cascading grid failure. The IAO origination might have any of a number of root causes, leading to a deviation between the rotor angles δ of different SGs emanating an electromechanical wave .e.g..:

- Wear and tear in the SG itself: rotor balance, erratic excitation
- Phase differences in the power flows between interconnected load balancing areas
- Weakness in the tie-lines - can span hundreds of kms - connecting different areas
- AGC mismatches between different areas
- Reactive power coupling, resonant frequencies

The greater the number of interconnections in AC grids, the higher the probability of IAO emerging. The US West Coast blackout of 1996 started from IAO of 0.1-2Hz, having to do with weak tie-lines between Canada and California.[14] The 2003 Italy grid blackout, though caused by human error, also involved power oscillations on the grid.[15] The IAO problem has been carefully monitored in China as well, in their grid expansion.[16] IAO waves from mismatched SG controls, can lead to cascading tripping. Grids suffering from chronic IAO resort to frequent load-shedding.

Inertia does not provide the grid with blanket immunity from voltage and frequency disturbances. On the contrary, excess inertia can make the grid too rigid for much needed speedy flexibility. Inertia can also be destabilizing,[17] e.g., in delaying responses to stabilize the grid, or temporarily masking a severe fault that grows into cascading tripping. A careful balance is needed, with more granular dynamic control, rather than blind faith in high inertia on the grid. No single type of power plant can meet the entire diversity of grid requirements. The grid requires elaborate measures, regulatory mechanisms, a carefully planned mix of resources to smoothly function for varying levels of inertia, flexible operational conditions, expanding interconnections, and timely responses to different types of disturbances.

Grid Codes

Clearly, a baseline set of technical rules is required for the management of HV, MV, and LV networks comprising the whole grid.[18] HV transmission and MV distribution networks are owned by grid operators, publicly or privately owned companies, while the LV network is in the ownership and responsibility of end-users. Grid operators issue grid codes governing the connection of electric power sources to their grids. Regulatory authorities maintain technical standards for the LV network, safety being of paramount importance, with grid operators specifying the rules for LV/MV connections including the use of DER (with the possible exception of when it is strictly for self-consumption, zero reverse power flows to the MV grid). The purpose is largely to prevent the types of problems we have been discussing, as what happens in one part of the grid affects another.

The rules are generally legally binding, subject to any special exceptions granted by the grid operator. For the larger power plants, the legal enforcement is strengthened through a grid connection agreement. Each power plant plan undergoes a grid impact study to assess it for grid code compliance. In order for electric power plant owners to plan their project development, how they might be able to integrate their plant into the grid, the grid code provides mandatory guidance.

Essential interrelated technical categories of a grid code cover fault management and safety, voltage and frequency regulation, with a long list of technical definitions, specifications, criteria in each category. Fault protection involves calculating variables like the short-circuit ratio, directional fault currents, specific voltage and frequency regulation responses including fault ride-through schemes. The specifications can vary by the size and type of power plant. Increasing sophistication is required in the design and implementation of transformers, switchgear, protective relays, and Supervisory Control and Data Acquisition (SCADA) systems. Voltage and frequency disturbances are continually occurring, of varying magnitudes, on a daily or weekly basis. A balance has to be maintained between protecting the grid equipment and unnecessary tripping of the generators, frequent load-shedding. Unexpected events will happen like trees falling on cables, rodents chewing a cable, bad connections, and unpredictable, extreme weather events. The objective of the grid code implementation is to prevent these disturbances from going out of control, ensuring smooth power delivery with minimal interruptions, and quick recovery from any lapses or accidents.

Power quality is another important concern. In AC grids, the fidelity of the power flow waveforms can get compromised by resonant harmonics and inter-harmonics. Non-linear circuit elements and reactive power compensation devices can cause complications. Managing this entire set of technical issues requires a comprehensive grid code, its diligent implementation, and close cooperation between power plant and grid operators.

Since most of the grids worldwide have been built for AC operation with SGs for the power sources, grid codes have been structured accordingly. The evolution of grid codes is deeply rooted in an AC electricity framework. An energy transition to new energy sources that produce DC power and loads like electric vehicles, data centres internally operating on DC, is a challenge for old school thinking. Demand response is a related topic of growing importance. So is cybersecurity in integrated grid controls to a more granular level. Incipient fault monitoring is crucial for stable operations of ever expanding, diverse, and interconnected grids. AC grids are here to stay, but an adaptation to greater DC networking is imperative. The future of electric power is both AC and DC, with DC gaining in application except when it comes to wireless power transfer.

--*--

Ch3 references, annotations, and supplementary reading

1. The theory and modelling of synchronous generators: Analysis of Electric Machinery and Drive Systems, 2013, Paul Krause et al. DOI: 10.1002/9781118524336
   Lecture notes on Synchronous Machines http://www.rakov.ece.ufl.edu/teaching/3211/Packet/5_Synchronous.pdf
   We have skipped the invention of induction generators that were initially used for AC electric power generation. Such generators are essentially asynchronous machines that cannot maintain stable frequency and voltage for a shared grid network. These machines also operate on the principle of electromagnetic induction, but the frequency and voltage are not separately controlled.
   "Slip Speed" is a key operational parameter of induction generators. The Slip Speed is the difference between two speeds: the speed of the magnetic field produced in the stator (the synchronous speed), and the physical speed of the rotor mechanical motion (e.g., driven by steam or the gravitational power of water). A negative Slip Speed is necessary for the delivery of active power by an induction generator - the reason it cannot set a synchronous frequency.
2. HVDC power transmission explained by a developer of DC power transmission and distribution systems, and an HVDC cable manufacturer:
   https://www.cencepower.com/blog-posts/hvdc-transmission-systems
   https://www.hitachienergy.com/us/en/products-and-solutions/hvdc/overhead-lines
3. A derivation of the voltage per phase for a synchronous generator:

https://circuitglobe.com/emf-equation-of-synchronous-generator.html

4. Swing equation, check the references in the link for more detail:
   https://en.wikipedia.org/wiki/Swing_equation
   For a fuller treatment of the topic:
   Power System Control and Stability, Chapter 2, IEEE Press, P.M. Anderson and A.A. Fouad
5. OCGT power plants in the UK operated by RWE:
   https://uk.rwe.com/locations/ocgt-power-plants-and-gas-engine-power-plants/
6. Mitsubishi Takasago CCGT:
   https://power.mhi.com/regions/amer/news/200402.html
   https://spectra.mhi.com/takasago-home-of-the-gas-turbines-at-the-heart-of-mhi-and-the-energy
7. Neurath coal:
   https://www.rwe.com/en/the-group/countries-and-locations/neurath-power-plant/
8. Flamanville 3 nuclear, France:
   https://en.wikipedia.org/wiki/Flamanville_Nuclear_Power_Plant
9. Grand Maison hydro, France:
   https://en.wikipedia.org/wiki/Grand%27Maison_Dam
10. US Federal Energy Regulatory Commission report on the 1977 NYC grid blackout
    https://www.ferc.gov/sites/default/files/2020-05/impact-77.pdf
11. US Department of Energy record of the August 2003 northeastern USA and southern Canada grid collapse:
    https://www.energy.gov/oe/august-2003-blackout

12. Subsynchronous Oscillation Problem Research in the UHVDC System of a Regional Power Grid in China, MATEC conferences, 2016, Ying Qu et al. https://www.matec-conferences.org/articles/matecconf/abs/2016/26/matecconf_mmme2016_04001/matecconf_mmme2016_04001.html
13. Union for the Coordination of Transmission of Electricity Report: The Effects of System Extension on Inter-Area Oscillations https://eepublicdownloads.entsoe.eu/clean-documents/pre2015/publications/ce/report_2002_6.pdf
14. Modes of Inter-Area Oscillation Oscillations in the Western Interconnection, 2021 report by Western Interconnection Modes Review Group report: highlights how IAOs led to the August 1996 grid blackout
15. Overview of the events and causes of the 2003 Italian blackout, Conference Paper 2016, Marino Sforna, TERNA Italian TSO DOI: 10.1109/PSCE.2006.29632
16. Analysis of inter-area oscillations in the South China Interconnected Power System, Electric Power Systems Research, 2004, Guoping Liu et al. https://doi.org/10.1016/j.epsr.2003.11.005
17. Grid Inertia: Good or Bad? David Houseman 2025 article: https://www.tdworld.com/reliability-and-resiliency/article/55303341/grid-inertia-good-or-bad
18. EU (ENTSO) grid codes: https://www.entsoe.eu/network_codes/

Ch4. Solar Cell

"Sol omnibus lucet" – Gaius Petronius Arbiter

Basic Physical Process

The thermal, chemical, and mechanical conversions to electrical energy that we have discussed, associated examples of power plants, do not directly convert the daily electromagnetic transfer from the Sun to useful available energy. Nature has known how to do it for a long time, that's why there is life on the planet, but our scientific engineering prowess has only recently mastered it. The breakthrough came from an understanding of semiconductor physics, based upon the theory of quantum mechanics.

A semiconductor is characterized by the energy bandgap E_g between its valence band (VB) of allowed electron energies, the uppermost filled level, and the next level of a conduction band (CB) where electrons can move as in a metal. An electron transition to the CB is said to leave a hole (an empty allowed energy state) behind in the VB: effectively a positive charge-carrier of equal and opposite charge to the electron. Electron-hole pairs, called excitons before their separation, are created. These VB-CB electron transitions can occur due to the exposure of semiconductors to photons, in what is known as the photovoltaic (PV) process.

At any non-zero temperature, electrons hop between the valence and conduction bands - so-called quantum leaps of random thermal energy - but without any net electrical effect. Photons from the Sun are absorbed, delivering electromagnetic energy to override the random thermal hops. The distribution of available energy states for electrons and holes is changed, in the CB/VB, by the formation of electron-hole pairs. This whole process is only explainable, and controllable, through our knowledge of quantum mechanics. Our control of semiconductor quantum mechanics is the cornerstone of modern electronics, including the computers and smartphones we regularly use.

A photon carrying electromagnetic energy $hc/\lambda \geq E_g$, absorbed in the first step of the PV process, induces an electron jump to the CB. The difference between E_g and hc/λ is dissipated as thermal energy by the photogenerated charge carriers in the semiconductor lattice. Photons with $hc/\lambda < E_g$ pass through the energy bandgap. An electrical potential difference, i.e., voltage, is established by E_g between the paired CB electrons and VB holes. Connecting the semiconductor, via electrodes, to an external circuit enables the movement of rapidly separated electrons and holes, collectively called charge carriers, to produce an electrical current. The combination of electrical voltage and current delivers DC electrical power, with a voltage proportional to E_g and current to the number of charge carriers that have been generated.

You may recall the solar spectrum of photon wavelengths in Ch1 (Fig.1) showing the irradiance distribution (yellow line) at the top of the atmosphere, and terrestrial photons after atmospheric filtering (red line) - the latter are available for the PV process.

Small E_g in a semiconductor allows a greater number of photons, i.e., more spectral wavelengths, to be absorbed to generate a greater number of charge carriers - for higher electrical current. A low E_g is thus attractive, but it also drives a lower electrical voltage due to smaller potential difference of the electron-hole pair between the CB and VB. Think of water flowing towards a wall, with the height of the wall analogous to E_g and the quantity of water flow to the number of photons. A taller wall results in higher potential energy of each water drop falling from it but also reduces the total quantity of water that can pass, and vice versa. Similarly, for a single semiconductor bandgap E_g there is an inherent physical trade-off between the V and I that can be generated from a given range of photons (the solar spectrum). Maximising V minimizes I, and vice versa. E_g must be engineered to balance V and I for maximum power conversion efficiency - the product of V and I. This trade-off is a limiting factor in PV energy conversion efficiency. It turns out that the ideal, single bandgap E_g to balance V and I for optimal PV conversion is 1.34eV.

The V, I trade-off is not the only factor to consider in making a useful energy generating cell from the semiconductor's E_g light absorption. Other vital considerations are rapid charge-carrier separation, the transport of the charge carriers to their respective electrodes avoiding recombination along the way, and charge-carrier extraction to the external circuit. The semiconductor material therefore has to be chosen not only for its E_g, but charge-carrier separation design, charge carrier transport properties, charge-carrier collection by compatible electrodes, and durability.

P-N Junction

The p-n junction was originally invented for the purpose of AC/DC conversion - a device called a rectifier. For many types of electric circuits, especially in sensitive electronics, DC is a necessity or highly preferable to AC. Before semiconductor technology, bulky vacuum tubes were used for rectifiers.

Semiconductors are "doped" with a controlled dose of impurities for an excess electrons or holes. In an undoped semiconductor, called an intrinsic semiconductor, the quantities of electrons and holes are equal. Crystalline silicon (c-Si) doped for excess holes, e.g., by a dose of gallium (Ga), becomes a p-type semiconductor labelled p-Si. Ga has one electron less in its VB than Si and VB electrons are used for chemical bonding so the introduction of Ga leaves Si with an excess of holes. Conversely, an "n" doped semiconductor, e.g., n-Si dosed by phosphorus (P), has an excess of electrons. Excess charge carriers of one type or the other enhance semiconductor conductivity - the charge carrier transport properties. For n-Si the majority and minority charge carriers are electrons and holes respectively - vice versa for p-Si.

When p-Si and n-Si are brought into contact the majority carriers, driven by the gradient of their differing concentrations, start diffusing to the other side. Diffusion continues till ionized dopants from the charge transfer on either side - e.g., negatively charged Ga in p-Si and positively charged P in n-Si - develop a built-in electric field of voltage V_{bi} that prevents further movement across a depletion layer barrier. Free charge carriers are pushed out of the depletion layer by V_{bi} to electrically divide the two sides.

The width of the depletion layer = $(2 \cdot \text{Permittivity} \cdot (V_{bi}/q) \cdot (1/(N_p) + 1/(N_n)))^{1/2}$.

Permittivity: degree to which the semiconductor material is polarized by an electric field

V_{bi}: the built-in voltage that depends on doping concentration

q: unit electron charge

N_p and N_n: respective concentrations of p- and n- dopants in the semiconductor

When a voltage V is applied in opposite polarity to V_{bi}, the width contracts to $(2 \cdot \text{Permittivity} \cdot ((V_{bi} - V)/q) \cdot (1/(N_p) + 1/(N_n)))^{1/2}$. Net voltage across the junction is reduced, narrowing the depletion layer. Depletion layers are on the order of 0.1 to 1 microns, depending upon the p-n doping profile.

Applying an external voltage against the V_{bi} to weaken it is called forward-biasing. The depletion layer contracts and electrical current flows through the p-n junction. An applied voltage in the other direction, called reverse-bias, expands the depletion layer to block current flow. Hence, the AC/DC rectification.

Sunlit Junction

The V_{bi} of an unbiased p-n junction suits the rapid charge carrier separation requirement of a solar cell. A semiconductor absorbs light and generates charge carriers as described earlier. V_{bi} selectively separates the charge carriers, pulling them by its electric field to their bulk majority sides. A pristine c-Si lattice supports high-mobility charge transport to the electrodes for collection. Electrodes are connected to the n-Si and p-Si regions, called bulk regions for majority carriers, to extract electrons and holes respectively - the electron collecting electrode called a cathode and the other an anode. Note electrical circuit convention shows the cathode (electron collector) as a negative terminal for a positive current direction opposite to the movement of electrons.

One of the electrodes facing the Sun is a transparent conductor typically a transparent conducting oxide (TCO) and or fine metal grid only, and the other can also be TCO or a metal – there is a metal grid on top of the front TCO except for all-back-contact designs. The game is to separate electron-hole pairs, then collect electrons and holes at the designated electrodes for each, while preventing the separated charge carriers from going to the wrong electrode.

Light absorption creates electron-hole pairs in the p-Si, depletion layer, and n-Si regions. Separated charge carriers in the depletion layer induce an internal photovoltage that weakens the V_{bi} of the p-n junction from within (forward-bias in the p-n junction diode rectifier does it externally to drive majority carriers). The internally lowered potential barrier facilitates the diffusion of separated minority charge carriers from either side across the depletion layer. Minority carriers switch sides to become part of the majority, drifting towards their electrodes for collection (drift superimposed on diffusion is driven by an electric field, in this case that of the resistive voltage drop in the majority bulk region when connected to a load). Light activated control of net voltage across the p-n junction paves the way for the minority carrier current driving the whole process.

One of the doped bulk regions, p- or n-side, is made thicker for sufficient light absorption and electron-hole pair generation – the thickness of the doped c-Si wafer (typically 245-295mm wafer diameters, trimmed to 130-150 microns thickness nowadays). Optical engineering for light trapping, i.e., increasing the optical length of photon travel in the cell, has reduced the thickness required for maximum light absorption.

The other side, the side facing the sun, is kept thin (typically 10-50nm) for higher transmission of light to the thicker bottom side but not too thin or its sheet resistance rises and so do recombination losses.

Depletion layer pairs:

Electron-hole pairs generated within the depletion layer are quickly separated and swept to one side by V_{bi}, with the electrons sent to the n-side and holes to the p-side. The role of the thin depletion layer is not primarily electron-hole pair creation, but separating the carriers through the V_{bi}. Minority carriers for the photocurrent are largely generated in the bulk doped regions.

p-side pairs:

An electron-hole pair generated on the p-side diffuses to the depletion layer, where it is separated by V_{bi}, with the minority electron migrating to the n-side for collection by its electrode (majority holes cannot switch sides because of the V-bi barrier). High-purity c-Si enables longer diffusion lengths to reach the depletion layer. When electrons reach the n-side electrode (cathode), assuming it is the TCO, they enter the TCO CB and drive electron flow into the external circuit for electrical current. Holes left behind on the p-side go to the metal electrode (anode), to meet incoming electrons. Total charge conservation, a fundamental rule of nature, is maintained across the circuit. Electrons move faster than holes in silicon, so at any instant there are more holes moving towards their electrode for total charge conservation.

n-side pairs:

The converse process takes place on the n-side to deliver current in the external circuit. The n-side holes of the charge carriers separated by V_{bi} are transferred across the depletion layer to the metal electrode at the p-side creating empty states (electron vacancies) in the metal. Electrons extracted from the TCO as electrical current in the external circuit, flow around the circuit to drop into the vacancies at the p-side metal electrode – an electron sink.

Electrical charges left behind on either side, by the minority charge carriers shuffled across the depletion layer, are balanced by the charge-carrier flow through the external circuit thereby delivering electrical power. Minority carriers are responsible for the photogenerated current. Majority electrons flow, from the n-side, in the external circuit to balance the internal current of minority carriers in the solar cell.

The electron-hole pair generation and separation is directly proportional to sunlight intensity. If the sunlight irradiance (measure of intensity in W/m^2) doubles, so do the electron-hole pairs that are generated and separated for the photocurrent.

All bases are covered: electron-hole pair generation, charge separation, transfer, and collection. These four PV steps are the basic design functions of a solar cell.[1]

Silicon p-n junction diodes have been used for AC/DC rectifiers since the 1940s - to date a bread and butter component of electronic circuits. All the materials processing and manufacturing technology developed for it, as well as that for silicon transistors and integrated circuits, benefits the production of silicon solar cells.

King Si

Over 95% of global solar cell production uses crystalline Si (c-Si) semiconductors.[2] The market share of c-Si has been increasing for the past seven decades, even more so after widespread PV commercialisation. There is a long list of compelling reasons for it that are not going to change anytime soon. Monocrystalline c-Si (mono c-Si) has predictably displaced poly/multi-crystalline silicon solar cells in the market for good. Made out of less pure crystals, cheaper poly/multi-crystalline Si solar cells were commercially popular for quite some time but the cost advantage faded with mono c-Si production economies of scale.

A solar cell that relies on a single semiconductor energy bandgap E_g is called single-junction i.e., a single junction for photon harvesting. The maximum power conversion efficiency of any material type of single-junction solar cell is 33.7%, corresponding to the optimal E_g of 1.34eV. This 33.7% efficiency is calculated for the incident terrestrial solar spectrum across all wavelengths at precisely defined measurement conditions. These measurement specifications are called Standard Test Conditions: STC of 1000W/m^2 irradiance, 298K cell junction temperature, AM1.5G spectrum. The STC condition of 1000W/m^2 is referred to as 1 Sun. The power conversion capacity of a solar cell at STC is called watt-peak (W_p) - the peak power capacity rating. Two theoretically possible exceptions to the 33.7% STC efficiency limit for single-junction solar cells, at 1 Sun, are hot electron and intermediate band designs.

In academic research, quantum efficiencies are sometimes quoted – internal quantum efficiency for the ratio of the number of electron-hole pairs collected to absorbed photons of a specific wavelength only. External quantum efficiency is the same ratio including optical losses of the photons that are not absorbed – but what practically matters is the total energy yield across the entire terrestrial solar spectrum (not just one wavelength) under actual operating conditions. Quantum efficiencies are important for research diagnostics.

The maximum single-junction efficiency has to account for radiative balance between the temperature of the Sun and that of the PV cell junction temperature, resulting in what is called the detailed balance limit of power conversion efficiency. Thermal radiative equilibrium between two bodies depends on their respective temperatures: the thermal energy being exchanged by radiation has to result in net zero thermal energy transfer. The solar cell junction is in thermal equilibrium with its surroundings and the Sun shining upon it, so it has to radiate some photons to maintain the net zero thermal energy transfer with the Sun (radiation is the only physical mechanism possible for it). This radiative process is the exact opposite of light absorption for electron-hole pair generation: electrons and holes combine to emit photons in what is called radiative recombination. The number of emitted photons and their energies depend on the junction E_g. A c-Si solar cell emits IR photons - E_g of 1.12 eV = 1100nm in the infrared - to maintain radiative equilibrium with the Sun and its surroundings. This unavoidable radiative recombination loss debits ~1.7% from the 33.7% single-junction maximum at STC. Therefore, the detailed balance limit efficiency of a c-Si solar cell at STC is reduced to 32%. There are two fundamental constraints determining the detailed balance limit: (1) electromagnetic energy matching between the solar spectrum and the cell junction (spectral energy optimization), and (2) net zero thermal energy transfer.

A third fundamental constraint of an unavoidable recombination loss involves three charge carriers, e.g., an interaction of two electrons and a hole, or two holes and an electron. Typically, two carriers recombine and the third has an inter-band transition. The third carrier dissipates its energy in the lattice as heat. The whole process is an inevitable outcome of the semiconductor electronic band structure. Debiting this three-carrier loss for c-Si, the maximum theoretical efficiency is reduced to a hard limit of 29.4%. The bottom of the CB and top of the VB are not aligned for c-Si in k-space (k stands for the allowed momentum states of an electron, fixed by the crystal lattice periodicity). The indirect bandgap weakens radiative recombination and three-carrier losses compared to a direct bandgap material. Three-carrier losses are relatively stronger because of semiconductor doping - the three-carrier interaction increases in a cubic dependence on free charge carrier concentrations.

To get close to the maximum theoretical c-Si limit of 29.4%, various types of power loss mechanisms have to be minimised.

Recombination with defects or impurities:

Losses from recombination of the electrons or holes in the bulk semiconductor occur because of defect traps and impurities, before the charge-carrier can make it to the electrode. It is the dominant loss in the bulk regions and depletion layer. Defects trap the carriers during or soon after separation. Other recombination losses are suppressed in the depletion layer by the V-bi effect.

The extremely high purity of mono c-Si preparation, with impurities less than a fraction of a part per billion to ten billion (9N, 10N pure c-Si), reduces bulk losses. Passivation, chemical and or electrical, is used to reduce losses from the edges of the depletion layer electric field where it abruptly terminates at surfaces. Field distortion and recombination traps are minimized.

Phonon scattering:

Phonons are quantized lattice vibrations, increasing with temperature. Phonon scattering reduces the charge carrier mobility, contributing to recombination losses by reducing the diffusion length and average recombination time.

Electron and hole mobilities are supported by the highly regular, ordered crystalline structure of Si over a long range with few defects.

Surface recombination:

Electron-hole pairs combine at the surface, due to surface defects and dangling bonds. Surface passivation techniques cut this loss. Passivation refers to chemically or electrically neutralizing defects:

chemical e.g. by saturating dangling chemical bonds with atoms, and electrical e.g. electrostatic fields to block charge carriers from recombining. For example, a combination of c-Si/SiO_2 is highly effective for surface passivation (a technique well known in silicon microelectronics).

Three-carrier recombination losses:

We mentioned how the process of electron and hole recombination transferring energy to a free carrier is unavoidable to a certain extent, but it is important to prevent an excess of it. Low doping for low charge-carrier densities minimises the three-carrier interaction. High purity silicon reduces the traps that can induce inter-band transitions to further curb this loss.

Excess radiative recombination:

Low doping densities keep excess radiative recombination to a negligible level, inhibited by the indirect energy bandgap of c-Si and a low radiative coefficient.

Electrical contacts:

Losses at the c-Si/TCO (HJT cells) or c-Si/metal grid front (PERC, TOPCon cells) and c-Si/metal back contacts - the TCO is typically Indium Tin Oxide (ITO) - can occur because of interfacial defects and work function mismatches (work function is a measure of the energy required to evacuate an electron from its surface energy state). The TCO is optimised for conductivity and optical transmittance. ITO/metal deposition techniques and doping for work function alignment cut these losses.

Electrical sheet resistance:

Sheet resistance losses arise from lateral conduction in the electrodes, due to the current flow in a two-dimensional planar device structure. The carriers flow vertically through the p-n junction to the electrodes and then laterally in the TCO and or a metal grid for current collection. Losses are reduced by the TCO doping, deposition technique for texture, and metal grid spacing design. Silver or copper is used for the metal grid.

Optical losses:

Anti-reflection thin film coatings minimise this loss. The high refractive index of c-Si enables total internal reflection in combination with lower refractive index layers like SiO_2. Light trapping in the cell design reduces optical losses - multiple passes of the photon through the photoactive layer to increase absorption, while avoiding thicknesses that may increase recombination loss mechanisms. All back contact designs further reduce these losses by removing the metal grid lines on the front transparent electrode.

Estimated STC efficiency penalties for a state-of-the-art heterojunction (HJT) solar cell:[3]

Recombination with defects, impurity traps: 0.1-0.4%

Phonon scattering contribution: 0.1-0.2%

Recombination of electron-hole pair with third charge-carrier: 0.1-0.3%

Surface recombination: 0.1-0.3%

Excess radiative recombination: < 0.1%, negligible

Electrical contact resistance: 0.1-0.2%

Sheet resistance: 0.1-0.2%

Optical losses: 1-2%

Total losses range: 1.6-3.5%

The control of these losses in the HJT solar cell design for industrial production has improved to 90-95% of the maximum theoretical efficiency. At the time of writing, the top reported STC energy conversion for a single-junction solar cell is a passivated back contact HJT design with 27.8% STC efficiency. TOPCon (Tunnel Oxide Passivated Contact) and PERC (Passivated Emitter Rear Contact) solar cell designs have reached nearly 27% and 25% respectively. PERC has been the industrial standard for the past decade, playing a key role in the commercial acceleration of PV, TOPCon overtaking it in the last four years, followed by HJT as the leading contemporary design. Combinations with interdigitated back contacts (all back contact cell designs) have led to significant performance boosts.

The indirect bandgap of c-Si is often mentioned as a reason for alternative thin-film PV research, but it isn't much of a problem. For a c-Si thickness of ~130 microns, there are sufficient phonons to provide a momentum kick for the indirect energy bandgap offset (lattice vibrations of 55meV at 298K) - over 95% light absorption at first pass. With the enhanced ability to control recombination losses in c-Si and advanced light trapping, the thickness is not a problem. The indirect bandgap actually helps in reducing radiative and third-carrier inter-band recombination losses. Photoactive layers of 130-200 microns are a small fraction of the total solar panel thickness - dominated by the glass (~2,000-4,000 microns) and cell polymer encapsulation (~800-1,000 microns) layers. Commercial thin-film and cadmium telluride (CdTe) solar panels generally require a greater thickness of glass and encapsulation for mechanical robustness and environmental protection. The durability of c-Si in outdoor operating environments is another major advantage of the material.

Solar cells were first used for space applications, to power satellites in the 1950s-60s. Some of these satellites are still in orbit. The p-n junction for solar cells consisted of a thicker p-layer (p-Si) ~200-300 microns for the light absorption with thin n-type (n-Si), ~0.1-0.5 microns in older designs, to form the p-n junction. The reason was that p-Si is more robust to background cosmic ray radiation. Consequently, the PV industry developed around p-Si ingots and wafers.

In recent years, the industry has shifted to n-Si because it enables higher performance solar cells (higher carrier mobility, purity of wafers with fewer defects) - 130 microns n-Si versus 160 microns p-Si for the light absorption in latest cell designs. Commercial solar cell designs, including PERC, TOPCon, and HJT are adaptable for both p-Si and n-Si as the light absorption layer.

Two-Diode Model equation for the PV I-V curve

Solar cell electrical performance, given by its I-V curve and power output, is a non-linear function of I and V (like that of a diode). The PV cell is modelled as a voltage-limited current source. Two diodes are connected in parallel on the model circuit for different paths of recombination losses. Series and shunt resistances are modelled for the resistive losses of bulk current flow and leakage current respectively.[4]

Figure 5: equivalent electrical circuit of a solar cell, showing series-resistance, shunt-resistance, and two diodes in parallel

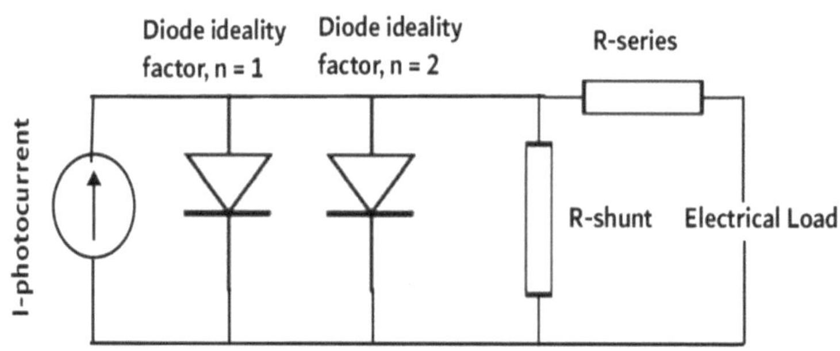

The two-diode I-V characterization of this circuit is given by

$I(V) = I_{ph} - I_1 \cdot (e^{(V+(I \cdot Rseries))/(n1 \cdot V\text{-}T)} - 1) - I_2 \cdot (e^{(V+(I \cdot Rseries))/(n2 \cdot V\text{-}T)} - 1) - (V+(I \cdot R_{series}))/R_{shunt}$

I (V): net circuit current

I_{ph}: photocurrent generated by absorbed light, proportional to light intensity

$I_{1:}$ diode 1 reverse saturation scaling for dark current leading to bulk recombination

V: voltage across the contact terminals

n1: a diode ideality factor of 1 for bulk region dark current

R_{series}: total resistance from the electrical contacts and bulk transport

V-T: thermal voltage = k (Entropy Constant) • temperature/unit electrical charge

$I_{2:}$ diode 2 reverse saturation scaling for dark current of carriers trapped in depletion layer aka space-charge recombination

n2: a diode ideality factor of 2 for depletion layer dark current

R_{shunt}: shunt resistance from leakage currents

The first term in the two-diode equation is the photocurrent proportional to absorbed sunlight (the cell insolation after optical losses). Bulk defects, surface, depletion layer traps, recombination losses in the charge-carrier transport, for a particular cell design, reduce the photocurrent. Energy bandgap E_g is reduced by higher temperature to increase the photocurrent (but the photovoltage decreases with temperature more than the increase in photocurrent, so there is a net decrease in power output with temperature rise).

The second and third terms I_1 and I_2 relate to the concept of "dark current". Subtraction of dark current, the sum of the two terms, results in the net electrical current from the PV current source.

Dark current flows in the p-n junction, even when it is not illuminated, through the thermal diffusion of majority charge carriers to the other side - becoming minority carriers - by overcoming the depletion layer built-in V_{bi} and recombining in the opposite bulk regions.

The diode ideality factors n1 and n2 are set equal to 1 and 2 to scale with eV/kT and eV/2kT respectively, corresponding to carrier concentrations in the bulk and depletion layers. For the depletion layer, the carriers are assumed to recombine in the middle of the layer's potential difference.

For a solar cell dominated by bulk recombination losses, I2 can be set equal to 0 - simplifies to the single-diode model for ease of computation. It is assumed that recombination losses in the depletion layer can be neglected or adjusted in n1 - in this case n1 varies between 1 and 2.

Reverse saturation measures the maximum dark current in thermal equilibrium in the absence of any voltage bias. Dark current increases exponentially with forward-bias voltage and temperature. The higher the E_g, the lower the dark current, and the greater the doping of majority carriers, the lower the dark current.

The final term in the equation measures power loss due to leakage current from the p-n junction, with the leakage current = $V_{junction}/R_{shunt}$, where $V_{junction}$ is the voltage across the p-n junction equal to $V + (I \cdot R_{series})$.

The numerator accounts for the voltage drop across the junction because of R_{series} - comprising bulk semiconductor, sheet, and contact resistances. R_{shunt} is for the parallel conduction paths of the leakage current, caused by manufacturing or material defects in the cell.

The open-circuit voltage V_{oc} is the maximum voltage, i.e., the induced photovoltage, of the solar cell when illuminated and no electrical current is extracted in the external circuit. It represents the total available energy potential between the CB electrons and VB holes: $q \cdot V_{oc}$ = the energy difference between the CB electrons and VB holes (the available free energy from the charge carrier chemical potentials shaped by incoming light). The theoretical limit of V_{oc} is set by E_g, even if there are no recombination losses ($q \cdot V_{oc} < q \cdot V_{bi} < E\text{-}g$).

V_{oc} decreases with rising cell junction temperature because of E_g temperature dependence and recombination losses. A linear dependence between V_{oc} and the cell junction temperature is derived, and used for calibrating the cell performance characteristics. At V_{oc}, the photocurrent and dark current cancel each other out (assuming there is no internal short-circuit leakage current).

In plotting I (y-axis) vs V (x-axis), V_{oc} is the x-axis intercept of the I-V curve.

The short-circuit current I_{sc} represents the maximum possible photocurrent and hence net current that can be extracted from the solar cell (the dark current terms go to zero), for a zero-voltage condition. Maximum electrons and holes are extracted with no opposing electric field (the opposing forward-bias to minority carrier current) and minimal recombination. I_{sc} increases in direct proportion to insolation. Plotting I (y-axis) vs V (x-axis), I_{sc} is the y-axis intercept of the I-V curve.

The I-V curve of a perfect solar cell has the shape of a rectangle with the maximum power point defined by P_{mpp}, maximum power output = $V_{mpp} \cdot I_{mpp}$, V_{mpp} the maximum voltage and I_{mpp} the maximum current at the maximum power point respectively.

A perfect cell requires zero series resistance, infinite shunt resistance, and negligible dark current (limited only by the radiative recombination of the detailed balance limit). For this ideal theoretical solar cell, impossible in reality, $V_{mpp} = V_{oc}$ and $I_{mpp} = I_{sc}$.

For a non-ideal solar cell, i.e., all real solar cells, the maximum power point will dip from the I_{sc}, V_{oc} values in the presence of non-zero R_{series}, R_{shunt}, dark current, and recombination losses.

These effects are combined in the I-V curve characterisation by a single metric called the Fill Factor (FF). The power conversion efficiency of an actual solar cell can be defined and calculated by V_{oc}, I_{sc}, and FF:

Power Conversion Efficiency (PCE) = $V_{oc} \cdot I_{sc} \cdot$ FF/Insolation (i.e. sunlight watts).

Figure 6: Conceptual I-V curve and FF for a real solar cell

https://en.m.wikipedia.org/wiki/File:Solar_Cell_I-V_Curve_Schematic.svg

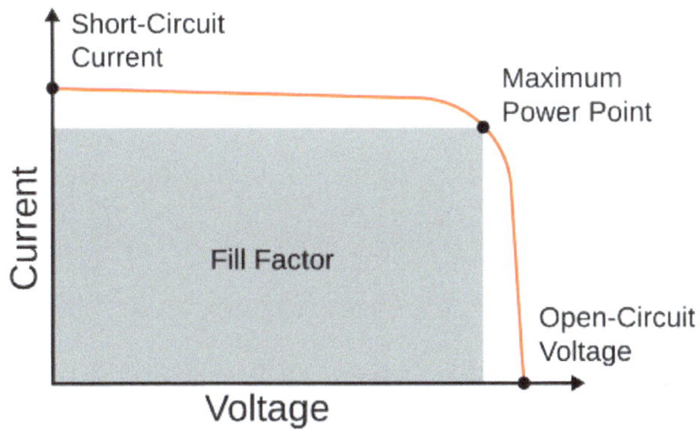

For an ideal solar cell (impossible in reality), the orange I-V curve line would be a perfect rectangle with $V_{mpp} = V_{oc}$, $I_{mpp} = I_{sc}$, in the diagram. The grey shared FF rectangle would completely fit the I-V curve. The closer the I-V curve is to a rectangular shape with V_{mpp} and I_{mpp} approaching V_{oc} and I_{sc} respectively, the higher the FF - ideal FF is 100%.

The two-diode equation does not have an analytical solution, so it is numerically solved for the I-V curves of different solar cell designs. The residuals of initial approximations of inputs to the equation are iteratively refined to converge to a fixed accuracy level. The numerical analysis is simplified for a one-diode model, assuming a constant R_{series} (independent of I, V), infinite shunt resistance, and I_2 set equal to zero.

Software packages are available for the numerical solutions. Industry standard commercial software PVSYST, PVSOL use one-diode models, SAM from NREL offers both one-diode and two-diode simulations.[5]

The shapes of measured I-V curves provide clues as to the types of defects and losses in a solar cell e.g.

- Tilted V slope near V_{oc} indicates high R_{series}, likely from poor contacts
- Tilted I slope near I_{sc}, indicating low R_{shunt}, hinting at poor encapsulation and wet leakage current
- Rounding of the curve near the maximum power point (MPP) indicating high recombination losses, possibly from interfacial defects and poor passivation
- Forward or backward bending near the MPP, a sign of light induced degradation
- Kinks in the curve - micro-cracks can cause it

The FF is reduced from unity (100%) by these anomalies in the I-V curve. High quality solar cells have smooth I-V curves with a smaller curved dip from a rectangle near the MPP. Top performing cells have achieved FFs exceeding 85%. Solar cell datasheets show the expected I-V curve under STC and normal operating conditions.

State of the art Silicon Heterojunction (HJT) Cell Design

A heterojunction is a p-n junction formed between two distinct semiconductors with different energy bandgaps (the E_g of one semiconductor has CB, VB offsets from the other), while a homojunction uses doped materials of the same semiconductor – as in PERC and TOPCon cells. Design principles overlap to a degree, so learning how one design works also enables an understanding of the others. The main motivation for HJT is that energy band offsets provide a powerful method to selectively separate and collect the charge carriers, thereby minimizing recombination losses.

HJT advantages vs PERC, TOPCon include:

- Superior a: SiH passivation vs SiO_2/poly-Si for TOPCon, PERC
- Higher V_{oc} > 750mV vs 700-730mV for TOPCon, PERC
- Energy barriers for carrier filtering vs doping gradients
- Low temperature fabrication ~470K vs ~1170K

The generic top-down architecture of an n-Si HJT cell is:[6,7,8]

- TCO metal grid (ITO with Ag): front transparent electrode
  CB: 4.5eV, VB: 8.0eV, work function: 4.7eV.
- Intrinsic amorphous Si (i-a-SiH) ~10nm: front passivation
  CB: 3.8eV, VB: 5.5eV
- Doped with amorphous Si (p-a-SiH) ~10nm: forms heterojunction with n-Si
  CB: 3.4eV, VB: 5.3eV

- n-Si ~150-200 microns: photoactive layer for electron-hole pairs

  CB: 4.05eV, VB: 5.17eV
- i-a-SiH ~10nm: rear passivation

  CB: 3.8eV, VB: 5.5eV
- n-a-SiH: electron selective layer ~10nm: heterojunction with n-Si

  CB: 3.5eV, VB: 5.3eV
- TCO with full area metallization: rear electrode

  CB: 4.2eV, VB: 8eV, work function: 4.0eV
- Dielectric-metal stack e.g. SiO_2/Ag for back reflection, light trapping

The illustrative eV values are absolute magnitudes for negative energy band levels going down from a vacuum reference level of 0 eV e.g., -5.3eV is deeper than -5.17eV.

HJT has two depletion layers at the front and back with energy band offsets and superior passivation, distinct from the other cell designs, for greater control over selective carrier collection.

VB and CB energy offsets on the front-side for selective hole collection are:

0.13eV VB p-a-SiH lower than n-Si VB – easing n-Si holes for anode

0.65eVCB p-a-SiH higher than n-Si CB – blocking n-Si electrons

Built-in electric field, voltage V-bi, points from n-Si to p-a-SiH: higher positive voltage at n-Si pushes holes to anode, electrons to cathode.

The p-a-SiH VB is below that of the n-Si by 0.13eV making it easier for n-Si minority holes to move towards the p-side electrode (anode), while majority holes from the p-Si are impeded from crossing into the n-Si.

The CB of the p-a-SiH is 0.65eV above that of the n-Si so majority electrons in the n-Si are impeded from crossing to the p-a-Si-H, while minority electrons from the p-a-SiH find it easier to cross over to n-Si for the n-side electrode (cathode) collection.

On the rear, the energy band offsets for electron collection are:

0.13eV VB n-a-SiH lower than VB n-Si, but low hole density in n-Si

0.55eV CB n-a-SiH higher than CB n-Si, high electron density in n-Si

Built-in electric field, voltage V-bi, points from n-a-SiH to n-Si: higher positive voltage at n-a-SiH driving electrons and holes to overcome respective energy offsets, electrons to the cathode and holes towards the anode.

High electron density in the n-Si and a thin intrinsic amorphous silicon layer i-a-SiH drives electron tunnelling to the cathode. A close alignment of the intrinsic and doped amorphous silicon energy bands supports passivation. i-a-SiH provides excellent surface passivation, in particular the saturation of dangling silicon chemical bonds with hydrogen, reducing surface recombination velocity to less than 5 cm/s vs > 100 cm/s for PERC.

The V_{oc} of 750mV is higher than other cell designs due to superior charge separation and passivation. Short-circuit current density is ~ 43mA/cm^2. TCO work functions at the front and rear are closely aligned to the doped a-Si energy bands to prevent a V_{oc} drop (0.1eV misalignment incurs a 10mV penalty), providing low electrical contact resistance and high carrier extraction at each end.

Recent innovations, e.g., the use of interdigitated back contact (IBC) technology, have raised HJT efficiencies to 27.8%. IBC places both the electron and hole collecting electrodes in an interlocking pattern at the back, with small gaps of 50-100 microns, eliminating front shading from a metal grid - there are only passivation layers and anti-reflection coatings at the front. Ultra-fast, precise laser ablation enables precise IBC patterning. Resistive and recombination losses are minimised by adjusting the contacts pattern for carrier mobilities. The use of nanocrystalline silicon in place of amorphous silicon and atomic layer deposition of alumina (Al_2O_3) for passivation are other performance boosting modifications.

HJT advantages extend beyond a higher STC efficiency:

- Lower temperature coefficient of power loss ≤ 0.25%/K
- Better performance for varying angles of incident light
- Longer guaranteed power performance lifetime of 30 years

- Lower annual power attenuation rate: < 1% the first year, ≤ 0.25% p.a. for year 2 to 30
- No LID (light induced degradation) and LeTID (light and elevated temperature induced degradation)
- Higher bifacial gain factor

These advantages become even more pronounced when the modules are mounted on trackers. As a result, the HJT lifecycle energy yield can exceed that of other solar cell designs by a margin of over 5%.

Bifacial Solar Cells

Bifacial PERC, TOPCon, and HJT solar cells are commercially available with transparent electrodes at the front and back for PV conversion from both sides – each side shares the same p-n junction V-bi for carrier separation. Excess doping in one layer creates an electric field, a "back surface field", to push the generated electron-hole pairs to the separation junction.

The semiconductor device is doped n^+-p-n^+, n-p-p^+ or p^+-n-p^+, p-n-n^+, depending on whether a p-type or n-Si wafer is used for the photoactive layer – plus sign superscript indicates excess doping. The n-type structure is more common in PERC, TOPCon designs because of better passivation of the n-type layers, improved charge-carrier mobility, and in HJT it suits the asymmetric energy offsets. Bifacial cells offer higher energy yields by over 5% vs monofacial panels of equal peak power capacity rating, when the albedo effect is significant, and are especially advantageous on trackers in ground-mounted projects.[9]

Cadmium Telluride (CdTe)

CdTe solar cells are second to Si in commercial usage. The compound semiconductor has an energy bandgap of 1.45eV, closer to the ideal single-junction E_g of 1.34eV for the terrestrial solar spectrum. As a result, it can theoretically achieve a higher PCE of 33% than 29% for c-Si under concentrated sunlight (multiples of 1 Sun irradiance: 10x for CdTe and 5x for Si – it has a higher thermal tolerance than c-Si for concentrated sunlight).

CdTe is combined with n-type CdS (E_g = 2.4eV) to form a deep depletion layer in the range of 0.5-2.5 microns. The n-CdS acts as a window layer allowing light to pass to the CdTe photoactive layer, generating the V_{bi} for charge separation. The direct energy bandgap, unlike Si, enables a thinner photoactive layer of ~2-5 microns. The temperature coefficient of power loss ~0.25%/K is similar to Si HJT, better than the other c-Si solar cell designs. A higher V_{oc} of 800-900mV versus 750mV for HJT is advantageous for the low-voltage (LV) circuit of a PV system – raising the voltage for lower current at the same power rating is beneficial. CdTe also has a spectral sensitivity edge with less of a performance drop for different angles of incidence, e.g., in early morning and late afternoon conditions. There is less of a difference between its PCE at STC and in actual operating conditions. The STC PCE is less than that of several PERC, TOPCon, HJT designs, but the gap narrows in actual operating conditions. The top CdTe cell STC efficiency is 23.1%, compared to 27.8% for Si. [10]

CdTe is competitive against Si for ground-mounted modules in hot conditions, with possible advantages in manufacturing cost and relative energy yield in actual operating conditions. It does not compete for rooftop modules, because solar panels made out of CdTe cells are thicker, heavier, with lower PCE for limited space.

Gallium Arsenide (GaAs)

GaAs has a direct energy bandgap of 1.42eV, closer than CdTe to the optimal single-junction E_g, enabling a photoactive layer of 1-2 microns. In addition, its crystalline structure offers much higher electron mobility: 5000-8,500cm^2/V•s vs 1,400cm^2/V•s for c-Si. STC PCE of 29.1%, FF 87%, has been achieved for heterojunction GaAs cells in a p-AlGaAs/n-GaAs structure (not to be confused with multi-junction cells using GaAs). Homojunction designs of p-GaAs doped with Zn and n-GaAs doped with Si have reached a PCE of 26%.

Under concentrated sunlight of 500 Suns, the efficiency of single-junction GaAs has reached 30.5%. With a hundred times greater radiation tolerance than c-Si, it can handle much higher levels of concentrated sunlight. Under concentrated sunlight, V_{oc} increases logarithmically and short-circuit current density J_{sc} rises to 1.25V and 15A/cm^2 respectively, to surpass the performance specs of other solar cells.

The drawback of GaAs for commercial use is expensive wafer material and manufacturing cost. GaAs crystals are grown using the liquid encapsulated Czochralski (Cz) process, vapour controlled Cz process, molecular beam epitaxy (MBE), or metal-organic chemical vapour deposition (MOCVD). These cost barriers may come down with advances in high throughput MBE/MOCVD equipment and shared materials demand for other power electronics devices. GaAs nanowires/c-Si tandem solar cells have also been investigated. In the meantime, GaAs is well suited to high-end space and defence applications.[11,12]

Perovskite

Over the past fifteen years, perovskites have emerged as the leading thin-film PV candidate superseding past interest in CIGS/CIS, polymer and small molecule bulk heterojunction solar cells. Perovskite refers to a chemical family of compounds ABX_3, where the A is for an organic (e.g. $CH_3NH_3^+$) or inorganic cation (e.g. Cs^+), B a metal cation (e.g. Pb^{2+}, Sn^{2+}), and X a halide anion (e.g. I^-, Br^-, Cl^-). Two frequently used perovskite compounds for solar cells are $MAPbI_3$ and $FAPbI_3$.

Perovskites were initially investigated for dye sensitized solar cells with TiO_2 scaffolding, but were quickly found to be more efficient in p-i-n and n-i-p thin-film device structures with bulk heterojunction variants of 2D/3D mixed phases. Perovskite solar cells (PSC) have a tunable direct bandgap E_g of 1.5-2.3eV, enabling photoactive layers thinner than 300nm. PSC efficiency may increase with temperature to an extent, and the design can be tuned for spectral sensitivity enhancement. Another attraction of PSC is its compatibility with low-temperature solution processing for cheaper manufacturing.

Single-junction PSC has reached 27.3% STC efficiency for an aperture area of ~0.05cm^2, with ~22% PCE for 1cm^2 aperture area (aperture area means the area exposed to full irradiance – it can be narrower than the full cell size for lab tests). Larger area efficiency is affected by inhomogeneities. The reported top efficiencies are rapidly changing, with larger area modules built out of PSC now being commercially marketed.

The device structures are more chemically complicated than p-n junction solar cells, with interfacial engineering of hole transport layers (HTL), electron transport layers (ETL), and interlinking molecules to prevent problems like ion diffusion. Since perovskites are water soluble, another issue is vulnerability to moisture ingress, as well as protection against outdoor operating conditions involving UV, heat, and humidity exposure. Hysteresis in the power conversion is also a challenge, being addressed by passivation methods for defects. Several approaches are being investigated to improve device stability e.g. completely inorganic perovskite cell designs. The most feasible commercial route for PSC is in tandem designs with c-Si.[13]

Hot Electron Solar Cell

Hot Electron Solar Cells (HESC) aim to harvest hot charge carriers, generated by photons with energies greater than E_g, before "thermalization" i.e., dissipating their excess energy in the lattice phonons. Thermalization takes place in less than a hundred femtoseconds. Calculations for HESC maximum theoretical efficiency under 1 Sun are in the range of 45-66%.

Hot carrier collection has been demonstrated in principle, but with very low efficiency as yet (single digits in special laboratory conditions). Nanostructured metal electrodes on semiconductors, targeting surface plasmons, have been tried. Quantum dot heterostructures, nanowires including tunnelling oxide wires, and asymmetric work function contacts for energy filters are being investigated. A critical challenge is simultaneously increasing optical absorption and efficiently collecting the generated hot charge carriers.[14]

Intermediate Band Solar Cells

Intermediate Band Solar Cells (IBSC) introduce a sub-bandgap to increase photon absorption, by absorbing lower energy photons than otherwise possible, without compromising the host E_g. Maximum theoretical efficiency of an IBSC under 1 Sun is 63% (possibly approaching 80% under concentrated sunlight, according to new calculations). The host E_g must be greater than 1.8eV (1.95eV for the 63% conceptual design calculation) for an intermediate band (IB) placement of a lower sub-bandgap by 0.7-1eV to the host VB and upper sub-bandgap by 0.7-1eV to the host CB. Sub-bandgaps have to add up to the host E_g to avoid V_{oc} loss. InAs/GaAs quantum dot (QD) IBSC with current constraint engineering has been certified for 22.8% efficiency, but under 93 Suns.[15] Current constraint engineering involves a top cell structure to limit the photocurrent, thereby masking V_{oc} and thermal losses – making it more of a tandem solar cell, rather than a true IBSC. Experimental attempts or ideas for a practical prototype include colloidal quantum dots, introducing deep level impurities in the bulk, highly mismatched alloys, and organic molecules for a sensitizer or high bandgap acceptor.[16] Stacked Si/SiGe/Si solar cells with graded energy bandgaps (multi-junctions) may offer clues to engineering a viable IB, or perhaps advances in metamaterials fabrication. The work on plasmonic solar cells might have some applicability to injecting hot carriers into an IB. Engineering the IB without defect traps is a formidable challenge. Truly ground-breaking experimental milestones have yet to be achieved but there is an ingenious theoretical framework to guide new efforts – IBSC has exciting potential for a dramatic breakthrough in solar cell efficiencies.

Multi-junctions

Multi-junction solar cells stack semiconductor materials with decreasing bandgaps from top to bottom, to harvest a greater portion of the terrestrial solar spectrum. Higher energy photons are successively captured for electric power conversion, with the lower energy photons passed down for their turn. The idea is to overcome the single-junction E_g trade-off between V_{oc} and I_{sc}. In theory, an infinite number of junctions can cover the entire spectrum absorbable by semiconductor bandgaps for a maximum theoretical efficiency of 68.2% under 1 Sun and 86.8% in concentrated sunlight (46,200 Suns). In practice, there are diminishing returns of energy yield in incremental junctions as there are fewer photon energies left to harvest. Over 95% of the maximum 68.2% efficiency is obtainable by 3-4 junctions. 37-40% cell efficiencies under 1 Sun have been accomplished by 3-4 junctions, 47.6% being the record for concentrated sunlight (500 Suns).[17,18,19]

Manufacturing complexities increase with additional junctions. The Bill Of Materials (BOM) for the semiconductors also disproportionately rises. Each additional junction must be current matched to the others, to avoid power losses. It becomes more difficult to find suitable semiconductors with the required bandgaps and crystalline structures for feasible lattice matching. Additional tunnelling junctions increase series resistance.

Practically, for commercial purposes, a dual-junction solar cell known as a tandem solar cell is the most attractive option. The maximum theoretical efficiency for a tandem solar cell with optimal bandgaps under 1 Sun is 45%, with a perovskite-silicon (P-Si) theoretical limit of 43% (perovskite at the top and silicon at the bottom).

A tandem solar cell is configurable in two terminals (2T), three terminals (3T), or four terminals (4T), all being investigated for commercial application:

2T:

A single positive and negative terminal is used, as in single-junction solar cells, for the power output. The top and bottom cell are connected in series with a tunnelling junction between the two cells for current matching. The advantage is simpler packaging as for regular solar cells. Disadvantages include precisely matching the bandgaps, the effect of insolation variation on performance, and PCE limitation by the lowest current cell.

3T:

A positive contact for the top cell, a negative contact for the bottom cell, and a shared third terminal between the positive and negative contacts, enables current to be independently collected from the top and bottom cells. It solves the problem of being constrained by the lower current cell, with a lower efficiency penalty than 2T. It is not as sensitive to sunlight fluctuations as 2T, and involves simpler manufacturing than 4T.

4T:

This configuration (essentially a mechanical stack) has fully independent terminals for each cell, with potentially the highest total efficiency since each cell can separately operate at its optimum MPP. The problems are that it involves higher manufacturing cost and is more difficult to integrate in a complete power delivery system.

P-Si tandem cells have been fabricated in the lab in 4T, 3T, and 2T configurations. The perovskite is tuned to a bandgap of 1.8-2eV for the top cell role. HJT is preferred for the bottom cell. Certified STC efficiencies of over 34% have been reported for P-Si cells of 1cm^2 area. The commercialization issues are cost, compatibility, stability, and durability in outdoor operating conditions.[20]

While almost everyone in the industry is focused on P-Si tandems, with Si at the bottom, an interesting alternative to consider is a tandem with Si HJT at the top and a lower E_g semiconductor for the bottom cell (e.g. investigation of germanium for this role). The tandem efficiency might be less than P-Si but possibly more durable with a longer lifetime, lower power attenuation rate, and higher energy yield to cost ratio.

Concentrated Photovoltaics (CPV)

CPV employs optical concentrators, parabolic reflectors, lenses, mirrors, to focus 100-1000x Suns on high-efficiency solar cells like the 37-40% multi-junctions. The CPV approach has failed to commercially prosper (only a few projects developed, with negligible market growth). One reason is the high cost of the optical equipment. Another is that it only uses direct normal irradiance (DNI) - the DNI ratio of total irradiance is reduced affected hazy atmospheric conditions - wasting the conversion capacity of diffuse light (DHI is the diffuse horizontal irradiance component).[21]

Si Cell Manufacturing

Manufacturing materials:

The materials utilization for mono c-Si cell manufacturing includes:

Wafers

- Wafers doped p- or n-type for the light emitting base (thick layer)
- Standard sizes: M10 182mm*182mm pseudo-square (square with rounded edges), M10R 182mm*199mm rectangle, G12 210mm*210mm complete square, G12R 182mm*210mm rectangle; the trend towards rectangular shapes is driven by space utilization for transport packing
- 150-180 microns thick for PERC, TOPCon
- 120-140 microns, thinner n-type for HJT
- 0.5-5Ω.cm resistivity

Front contact dopant

- Phosphorus Oxychloride $POCl_3$(P) for PERC n-type emitter, nitrogen (N_2), oxygen (O_2) carrier gases for chemical reaction
- Boron Tribromide (BBr_3) for TOPCon p-type emitter, nitrogen (N_2), oxygen (O_2) carrier gases
- Phosphine (PH_3), silane (SiH_4) for silicon source, hydrogen (H_2) carrier gas, for doped amorphous silicon layer in HJT rear emitter design
- Diborane (B_2H_6) or trimethyl boron (TMB), for HJT front emitter design

Back contact dopant

- Aluminium paste (p^+), PERC
- Poly-Si tunnel oxide deposition n-Poly-Si(P): SiH_4 for poly-Si deposition with PH_3 for phosphorous (P) dopant and H_2, N_2 carriers; nitric acid (HNO_3), hydrogen peroxide (H_2O_2), or ozonated deionised water (O_3-DI) for tunnel oxide wet chemical growth, oxygen (O_2) for thermal, TOPCon
- Phosphine (PH_3), SiH_4 and H_2 carrier for hydrogenation for front-emitter cell, B_2H_6 or TMB, SiH_4 and H_2 for rear emitter

Passivation layers

- SiH_4, ammonia (NH_3), N_2 carrier, for silicon nitride (SiNx) PERC front, trimethyl aluminium (TMA) and H_2O or ozone

(O_3) for rear alumina (Al_2O_3), SiH_4 and NH_3 for rear SiNx or silicon oxynitride (SiOxNy)

- SiH_4, NH_3, for TOPCon front, with TMA for Al_2O_3 option, O_2 for thermal growth of silicon dioxide (SiO_2) tunnel oxide rear, HNO_3 and O_3 for SiO_2 chemical growth, SiH_4 and NH_3 for SiNx rear capping
- SiH_4 and H_2, HJT front and rear

Contacts metallization

- Ag or Ag/Al paste for the high temperature firing of TOP front and back contacts, Ag for PERC front, Al at PERC back
- Nitrogen (N_2) or forming gas (N_2/H_2) for furnace atmosphere, for PERC and TOPCon
- Low temperature Ag paste for HJT

Texturing

- Potassium Hydroxide (KOH) or sodium hydroxide (NaOH) with isopropyl alcohol (IPA) surfactant for alkaline etching, standard for all cell types – various new surfactants, additives are substituting or reducing IPA usage
- NH_4OH and H_2O_2 for inverted pyramids (etched into wafer), Nitric Acid (HNO_3) polishing or rounding of pyramids, and tetramethyl ammonium hydroxide (TMAH) as a high-quality alternative to KOH for HJT

- Sulfur hexafluoride (SF_6) and O_2, for plasma reactive etching of “black silicon” (not required for high efficiency mono-Si solar cell designs)

Cleaning

- Wet cleaning chemicals set from the silicon microelectronics industry (named after RCA, Radio Corporation of America):
  Alkaline peroxide mixture (NH_4OH, H_2O_2, H_2O), for organic residues, hydrochloric peroxide mixture (HCl, H_2O_2, H_2O), metal-ion contaminants, for all cells, and HF dip to strip away oxide layers (diluted more to 1-2% HF for TOPCon, HJT)
- Alkaline peroxide mixture being largely replaced by ozone-based methods
- Mixture of sulfuric acid (H_2SO_4) and H_2O_2, post-texturing for PERC, TOPCon
- HCl-HF for phosphosilicate glass (PSG) removal, O_3 for oxide growth, TOPCon
- Hot O_3-DI, organic acids (citric, oxalic), HJT

Protection

- SF_6 or tetrafluoromethane (CF4) mixed with O_2, edge isolation etching, or HF, HNO_3 for edge stripping
- Photoresists or organic coatings for patterning masks (e.g. Polymethyl Methacrylate, PMMA), acetone or isopropyl for mask removal

- Thin plastic films (PE, PET) or acid-free, anti-static, paper separators for storing cells in trays

Fabrication steps for PERC (p-type), TOPCon (n-type), HJT (n-type) designs:

An outline of the manufacturing process sequence - in practice, it is not a linear workflow but structured in batch processes to maximise the wafer throughput per hour:

Czochralski (Cz) process for pulling high purity monocrystalline ingots - up to 300mm diameter

- n/p-type dopants are added to melt: Gallium (Ga) for p-type, Phosphorus (P) for n-type
- Cone ends of the pulled Si crystal are chopped off for a uniform cylinder
- Specialized diamond-plated band saw blades or diamond multi-wires in a squaring machine trim the cylinder into a square or rectangular ingot

Ingot slicing

Diamond wire saws dominate the slicing of Cz ingots into wafers.[22]

Kerf loss in ingot slicing:

- As much as 40% kerf loss of ingot for PERC, TOPCon
- 35% for HJT: thinner wafers with tighter control of total thickness variation
- KOH chemical bath uniformly etches away the surface for saw damages, scratches, and rough edges.

- Polyethylene glycol (PEG) and or deionised water cools the diamond wire saw cutting
- PEG/H_2O slurry of swarf (kerf loss) is washed away, and the rinsed wafer dried by compressed N_2.

Wafer cleaning and pre-texturing

- After the KOH bath, the wafer undergoes wet chemical cleaning by immersion in a hot mixture of NH_4OH, H_2O_2, and DI-H_2O, (deionised water) to remove organic contaminants, followed by HCl, H_2O_2, and DI-H_2O for removing metals, and then a HF dip to leave a hydrophobic, hydrogen-terminated surface
- DI-H_2O rinse and drying (e.g. spin dryer, or IPA vapour dryer) in the final preparation step for texturing
- Plasma enhanced dry cleaning using reactive species of oxygen (O_2), hydrogen (H_2), ammonia (NH_3) plasma (charged particles) – avoids the use of wet chemicals, but uniform cleaning across the wafer is a challenge, for HJT, TOPCon

Texturing

The texturing is primarily for light trapping by creating a vast array of random pyramids on the wafer surface:

- Alkaline etching, using KOH, controlled by high boiling point surfactants or additives, is the low-cost standard

More advanced, premium options:

- TMAH wet etch, slower and more controllable than KOH,

- SF_6/O_2 plasma reactive ion etching, followed by HNO_3 polishing,

Wafer cleaning, post-texturing

PERC, TOPCon:

- Immersion in a sulfuric-peroxide mixture (called "Piranha") to oxidize and remove organic contaminants, followed by a DI-H_2O rinse
- Bath of ozonated deionised water, O_3/DI-H_2O to remove metals, particle contaminants
- Megasonic deep cleaning, and HF-O_3 for PERC
- Ultrafine O_2 or N_2 gas bubbles, dissolved in DI-H_2O bath – finer bubbles for TOPCon
- HF dip - diluter concentration for TOPCon
- HCl-O_3 (TOPCon) for controlled oxide growth ~1.2-1.5nm
- IPA vapour and an N_2 stream in the "Marangoni" process, creates a surface tension gradient for drying

HJT:

- Organic acids cleaning and hot DI-H_2O rinse
- Dilute HF (1-2%) dip, followed by final DI-H_2O/O_3 rinse
- Immediate transfer to vacuum lock of PECVD, to prevent native oxide growth even a few angstroms thick, after the surface tension gradient drying

Doping and Passivation

In these two steps, the p-n junction is formed, electrochemically passivated, and sealed by energy-loss barriers. The thinner side of the p-n junction is called the emitter.

PERC:

Doping by phosphorus diffusion

- N_2 carrier gas from liquid phosphorus oxychloride ($POCl_3$) passes into a quartz tube holding the wafers at 1100-1200K, creating PSG glass layers on the wafer surfaces
- At the high temperature, phosphorus atoms diffuse into the p-type silicon wafers for n-type emitters, forming a p-n junction

Rear passivation

- After etching of the front PSG and rear, a thin ~10nm Al_2O_3 layer is assembled by atomic layer deposition (ALD), with precursors of TMA and an oxidant H_2O or O_3, to fix dangling chemical bonds for passivation
- Thicker SiNx capping layers are deposited by plasma enhanced chemical vapour deposition (PECVD) to provide mechanical support and field-effect passivation

Front passivation

- PECVD of SiNx for field-effect passivation and anti-reflection coating (ARC)

TOPCon:

Doping and rear passivation

- Liquid BBr_3 is carried by N_2 into a high temperature furnace >1220K, holding the wafers, to form a borosilicate glass (BSG) layer with high temperature crystallization forming the front p-type emitter, wraparound BSG is etched away by 1-2% HF
- A poly-Si layer is deposited by low pressure chemical vapour deposition (LPCVD), or PECVD, on the ~1.5nm SiO_2 over the wafer surface with phosphine (PH_3) and SiH_4 for in-situ doping; high temperature crystallization drives n-dopants into the poly-Si interface to form a p-n junction and a passivating rear contact

Front passivation

- PECVD of Al_2O_3/SiNx for front passivation and ARC

HJT:

Passivation

- PECVD of ~5-10nm layers of i-a-Si:H on the front and rear at a relatively low temperature of <475K to saturate the wafer surface's dangling bonds

Doping

- PECVD of p-type front, with SiH_4 and p-type dopant B_2H_6 or TMB, and n-type rear, with SiH_4 and n-type dopant PH_3, forming asymmetric p-n junctions

- Passivation and doping are in the reverse order for HJT - unlike PERC or TOPCon, it does not require high temperature processing.

TCO deposition process for HJT only

- Sputtering for precisely controlled deposition
- Argon ions bombard the ITO target, with O_2 introduced to chemically react and control the properties of the ITO contact formation
- Lasers open the TCO contacts by precisely removing the intermediate layers for silver printed or copper plated contacts to the underlying silicon
- Pulsed lasers are used to remove shunts between the front and back contacts, ablating conductive paths at the edges

Lasers or alternatives for patterning

- Laser ablation of openings in back contact for PERC, to form a back surface field during metallization
- 2-3 steps of laser scanning (laser selection emitter technology) for localized heating of the TOPCon cell areas where silver contacts will be metallized - higher doping concentrations for low resistance contacts and low doping concentrations in between the contacts
- Patterning of contact areas by shadows, masks for HJT

Metallization

Metallization of fingers and busbars on the cells, for interconnection in the module assembly process. The fingers and busbars (BBs) are screen printed in a single step, and then fired together for PERC, TOPCon or cured in a low temperature process for HJT.

- Screen printing Ag fingers, BBs, front and rear, or full area Al with Ag pads on the rear, with high temperature firing ~1170K, for PERC and TOPCon
- Screen printing and low temperature infrared (IR) curing at ~470K lower temperature, for HJT, Ag fingers BBs front and/or rear

The printing of the metallization is done separately for the front and rear, and then the firing (PERC, TOPCon) for both sides is done at the same time.

Copper (Cu) metallization to replace Ag, at both the front and rear, is an ongoing industry trend – application of Cu coated Ag paste has been implemented in industrial manufacturing.

Screen-printing of Cu in lab HJT cells, and Cu electroplating by a leading manufacturer for commercially deployed solar cells has been implemented to completely replace Ag.

In back-contact designs, Cu electroplating is only required on one side. For bifacial cells, light induced Cu deposition is a well-known method that can be used for the front. Cu ions are toxic to Si so a thin intermediate layer of nickel silicide (NiSi) is sintered. The challenge of delicate wafer handling for Cu deposition has been solved.

Lasers for dicing wafers into individual cells

Lasers have largely replaced diamond wire saws for wafer dicing; higher throughput and well-suited to half-cut cells among other new designs[23]:

- Laser dicing has 2-3x higher capex cost, but reduces waste
- 5-15% wafer waste vs 30-40% for diamond wires
- Higher precision and speed, non-contact cutting:
- 6,000-10,000 wafers/hour vs 2,000-4,000 wafers/hour by diamond wires
- Cleaner cuts with fewer micro-cracks at edges, though diamond wires also have acceptable good quality
- Precision for half-cut cells and cells for shingled modules
- Superior for thinner wafer thicknesses < 160 microns
- N_2 assists laser dicing by blowing away vaporized silicon in the process

Testing

- Electroluminescence (EL), Photoluminescence (PL) for defects inspection
- Flash testing, for I-V curves
- More granular inspection includes infrared imaging, lock-in thermography (LIT), dark I-V testing

Packaging

- Primary containment: enclosed in static-free plastic carrier cases, e.g. polypropylene (PP)

- Surface protection: thin sheets of paper between the cells to prevent scratching
- Environmental protection: carrier cases are sealed in anti-static, cleanroom certified bags (e.g. nylon or metallized PET) to protect against oxidation of contacts and electrostatics

The actual long-term performance of a PV system is rooted in the quality of these cell manufacturing materials and processing steps.

Bill-Of-Materials (BOM)

The BOM accounts for most of the total production cost, especially at high production throughput: 75-80% for PERC, 85-90% for TOPCon because of higher Ag consumption, 80-85% for HJT. The rest is around 3-5% for energy, 4-7% equipment maintenance, depreciation (higher for HJT), and labour (not much required because of automated production lines).

About 60-70% of the BOM cost is that of the high-quality c-Si wafer, with n-type 5-8% more expensive. Silver (Ag) is the second most expensive item in the BOM: 12-15% for PERC, 18-22% for TOPCon, 15-18% for HJT. There is rapid ongoing progress in reducing Ag consumption for all three cell designs, including copper substitution for the metallization (e.g. by Cu electroplating). Recently, Ag consumption has been minimized for HJT designs:

HJT 0-5mg Ag/W_p < PERC 6mg Ag/W_p < TOPCon 7-9mg Ag/W_p

The rest of the BOM: 15-20% PERC, 15-20% TOPCon, 18-22% HJT covers the electrodes, metal grids, passivation layers, and overheads. Poly-Si is the main distinguishing cost between PERC and TOPCon.

Polysilicon

Solar-grade polysilicon, a high purity form of polysilicon, is the primary feedstock material for silicon solar cells. Latest purity levels range from 9N to 10N (10N means a purity percentage to eight decimals after 99% or one in billion impurity level i.e., 99.99999999%). Metallurgical grade silicon (MG-Si) is the precursor to polysilicon – processed out of quartz mining. Silicon is the second most abundant element in the Earth's crust after oxygen, so there is no risk of running out of it. The material is also completely recyclable. Indeed, in the early days, the PV manufacturing supply-chain used silicon scraps from the microelectronics industry. Before 2000, over 90% of polysilicon was used for the microelectronics industry. Now in 2025, it is less than 5% for microelectronics with the rest for solar.

The industrial processes for polysilicon production from MG-Si have been known for decades - essentially the same processes are being used today with efficiency improvements over the years. The main methods use trichlorosilane, dichlorosilane, or silane in a hot reactor. Fluidized bed reactors are another production route. About 80% of the market uses the trichlorosilane process in a

CVD reactor method, with an energy consumption of 50-80kWh per kg of polysilicon. Polysilicon production plants are very capital intensive ~$1-1.5 billion required for 100,000 tonnes per year. It was therefore presumed that polysilicon prices might follow a cyclical pattern but for the last three to four years, there has been a glut of polysilicon with record low prices of $4-$5/kg.

Monocrystalline silicon ingots are produced from polysilicon using the Cz process: 1.2-1.5kg of polysilicon to produce a kg of mono-Si ingot, consuming 25-40 kWh/kg. Then about 12-15g of the ingot used for a single M10 wafer, 16-20g ingot for a G12 wafer. G12 wafers of 130 microns weigh about 13.5g, for ~1.6g ingot c-Si/Wp and ~1.2g wafer c-Si/Wp respectively. The energy consumed by the diamond wire saw and lasers for the ingot slicing and wafer cutting is in the range of 1.5-3kWh/kg.

At the time of writing, global annual polysilicon production is 1.7-2 million tonnes per year, with reported inventories of 300,000-400,000 tonnes. Polysilicon, wafer, cell, and panel prices are weekly updated and publicly available.[24, 25, 26]

High-Volume Throughput

The production throughputs for high-volume manufacturing lines are in the vicinity of 125MW$_p$/week for PERC, 100MW$_p$/week for TOPCon, and 65MW$_p$/week for HJT (estimates for M10).[27]

The rate limiting steps of the manufacturing throughput vary by cell design:

PERC:

The rate limiting and most energy intensive manufacturing process for PERC is the wafer doping by $POCl_3$ diffusion furnaces ~0.4kWh/m^2 (wafer area, ~15kWh/m^2 where 1 m^2 is the cell area), 500-700 wafers per hour in 3-6 batches. It is for the n-dopant in the p-n junction of a p-Si solar cell. The process requires a furnace temperature of ~20-30 minutes at 1120-1170K, 10-20 minutes for batch cooling, heating ramp, loading, and unloading time.

The process is slow, but simple, using low capex equipment, and relatively cheap operating cost. For selective emitter PERC cells (SE-PERC), laser or ion beam doping is used. Laser doping and ion beam implantation are widely used techniques in microelectronics, the integrated circuits industry.

TOPCon:

TOPCon also uses $POCl_3$ diffusion furnaces - the most time consuming and energy intensive manufacturing process once again, with a similar throughput time per wafer and energy consumption per wafer. An additional processing step of Poly-Si ~1,800-2,200 wafers per hour is inserted in the total cell throughput time.

HJT:

The PECVD of intrinsic amorphous silicon (a-SiH) deposition for passivation is the most time consuming and throughput rate limiting manufacturing process ~8-10kWh/m^2, cell area, 1,200-1,600 wafers per hour. ITO sputtering of the TCO is also one of the more time-consuming processes: 1,500-2,000 wafers per hour. The total throughput time of HJT is slower than PERC and TOPCon because it has more sequential processing than the parallel batch processing workflows of the latter. Every wafer has to go through the PECVD process of i-a-SiH deposition sequentially.

The legacy PERC lines are easily extendible to TOPCon by adding new tools for ALD, poly-Si deposition, and laser doping. HJT production requires the installation of a completely new manufacturing line. Capex for 65MW/week HJT manufacturing capacity is in the ballpark of $100 million.

Solar cells are not of an ideal size for end-consumer use, as the devices generate high currents at relatively low voltages. Power sources with a higher voltage to current ratio, i.e. less current per watt, are preferable - lower currents reduce resistive power losses, enable easier and more economical system integration, and are generally safer to plan (to an extent). Therefore, cells are interconnected in series and parallel to produce commercial solar panels. The solar cell is the energy generating core of a PV system, but its final commercial application depends on solar panel design.

--*--

Ch4 references, annotations, and supplementary reading

1. Physics of Solar Cells: from Basic Principles to Advanced Concepts, published by John Wiley and Sons, Peter Wurfel and Uli Wurfel. DOI:10.1002/9783527618545, details the basic physics of solar cells.

   The probability distribution function f(E) that an electron occupies a particular energy state as a function of temperature is called the Fermi-Dirac distribution: f = 1/(1+e(E-E_f)/kT), where E_f is the equilibrium chemical potential (recall the available energy functions in Ch2) called the Fermi level, k is what we called the Entropy Constant. The probability of electron occupancy is ½ at the Fermi level, i.e., $f(E_f) = 1/(1 + e^0) = ½$. Under illumination, the quasi-Fermi non-equilibrium distributions of electrons and holes emerge:

   Electrons: $f_n(E) = 1/(1 + e^{(E - Efn)/kT})$

   Holes: $f_p(E) = 1/(1 + e^{(E - Efp)/kT})$

   Photovoltage $q \cdot V_{oc} = Ef_n - Ef_p$

   Ef_n rises towards the CB, and Ef_p falls towards the VB. The emergence of the two non-equilibrium distributions for electrons and holes, under illumination, is called Fermi level splitting.

   The various recombination loss mechanisms are discussed in detail by Wurfel et al. SRH recombination refers to recombination losses due to trap states of defects and impurities. Auger recombination is the standard term for recombination losses involving a third free charge-carrier.

2. Twenty years of silicon photovoltaics: progress and future challenges,

   Volume 32, Issue 5, M.A. Green.

   By the inventor of the PERC solar cell design, the workhorse of the PV industry for many years.

3. Highest STC efficiency solar cell reported to date of 27.81% silicon heterojunction (HJT) interdigitated back contact (IBC) design ~ 1.6% less than maximum theoretical efficiency silicon single-junction: https://www.longi.com/eu/news/dual-world-record-hibc-tandem/
4. The online site pveducation.org covers the double-diode model for solar cells and the diode ideality factor, among a comprehensive menu of PV topics: https://www.pveducation.org/pvcdrom/characterisation/double-diode-model
5. PVSYST, PVSOL, SAM with PVWatts models, Aurora Solar software simulation packages, among others:
https://www.pvsyst.com/
https://pvsol.software/en/
https://sam.nrel.gov/
www.aurorasolar.com
6. The Physics of Amorphous Silicon/Crystalline Silicon Heterojunction Solar Cells, Journal of Applied Physics, 2012, DOI: 10.1063/1.3106642, H. Fujiwara and M. Kondo
7. Silicon heterojunction solar cells with up to 26.81% efficiency achieved by electrically optimized nanocrystalline-silicon hole contact layers, Nature Energy, 2023, Hao Lin et al. https://doi.org/10.1038/s41560-023-01255-2
8. 27.09%-efficiency silicon heterojunction back contact solar cell and going beyond, Nature Communications, 2024, DOI:10.1038/s41467-024-46726-5, Z. Cai et al.
Bifacial solar photovoltaics – a technology review, Renewable and Sustainable Energy Reviews, 2016, R. Guerrero-Lemus et al.
DOI: 10.1016/j.rser.2016.03.041

9. First Solar is the leading industrial manufacturer of CdTe solar cells and modules, with a knowledge centre of extensive technical documentation.
https://www.firstsolar.com/Resources/Downloads

10. High-Efficiency GaAs-Based Solar Cells, 2020, Masafumi Yamaguchi
DOI: 10.5772/intechopen.94365

11. Modeling and design of III-V heterojunction solar cells for enhanced performance, Cell Reports Physical Science, 2023, Kevin Schulte et al.
https://www.cell.com/cell-reports-physical-science/fulltext/S2666-3864(23)00338-7

12. A comprehensive review on the advancements and challenges in perovskite solar cell technology, RSC Advances, Volume 14, Issue 8, 2024, M.A. Nazeeruddin et al.
DOI: 10.1039/D3RA07518D

13. Hot electron plasmon-protected solar cell, Journal: Optics Express, Vol. 23, No. 15, 2015, J. Kong et al.
DOI: 10.1364/OE.23.0A1087

14. High-efficiency InAs/GaAs quantum dot intermediate band solar cell achieved through current constraint engineering, Next Materials, 2023, Thomas Sugabe et al.
https://doi.org/10.1016/j.nxmate.2023.100013

15. Intermediate band solar cells: Present and future, Progress in Photovoltaics: Research and Applications, 2020, Inigo Ramiro and Antonio Marti
DOI:10.1002/pip.3351,

16. Six-junction III–V solar cells with 47.1% conversion efficiency under 143 Suns concentration, Nature Energy, 2020, John F. Geisz et al.
DOI: https://doi.org/10.1038/s41560-020-0598-5
17. Overview and loss analysis of III–V single-junction and multi-junction solar cells, EPJ Photovoltaics 2022, Yamaguchi et al.
DOI: 10.1051/epjpv/2022020
18. Advances in multijunction solar cells: an overview, arXiv, 2024, Anika Tabassum Raisa et al.
DOI: 10.48550/arXiv.2405.01550
19. A Review on Perovskite/Silicon Tandem Solar Cells: Current Status and Future Challenges, energies, 2025, Jingyu Huang and Lin Mao
https://doi.org/10.3390/en18164327
20. A Review of Solar Photovoltaic Concentrators, 2014, International Journal of Photoenergy,2014, Mehrdad Khamooshi et al.
DOI: 10.1155/2014/958521
21. Laser cutting specialist for PV:
https://scantechlaser.com/solutions/photovoltaics/
Explains the applications and advantages of laser cutting
22. Diamond wire saw cutting specialist:
https://www.ensolltools.com/the-complete-guide-to-monocrystalline-silicon-wafer-cutting-diamond-wire-saw-cutting-process/
23. International Technology Roadmap for PV (ITRPV), 2024
Materials usage and manufacturing trends
https://www.qualenergia.it/wp-content/uploads/2024/06/ITRPV-15th-Edition-2024-2.pdf
24. Bernreuter Research – polysilicon market expert:
https://www.bernreuter.com/polysilicon/

25. Polysilicon, Si wafers, PV cell, module price tracker: https://www.trendforce.com/price/pv/polysilicon

26. Shipments by top five solar cell producers for H1 2025: https://en.tongwei.com/news/detail/168963.html
87,500MW solar cells shipped in the first six months of 2025 from the top five manufacturers by volume namely Tongwei, Runergy, Yingfa Energy, Jietai Technology, and Aiko Energy. Tongwei has reportedly crossed an annual cell production capacity of 150GW.

Ch5. Solar Panel

"Même la nuit la plus sombre prendra fin, et le soleil se lèvera" – Victor Hugo

The purpose of the solar panel, aka the PV module, is to integrate and package solar cells for commercial application. PV modules have become a globally traded commodity of energy, similar to a barrel of oil. A consumer cannot keep a barrel of oil for long if they want to extract energy from it at the same time, but that is not the case for a solar panel. The principal difference being the rate at which sunlight is converted to useful available energy in oil and a PV module - it takes millions of years of catagenesis for oil, but with the module it is instantaneous.

Topmost objectives of the cell-to-module (CTM) fabrication process are delivering a minimal gap between cell and module efficiency, for minimal added cost, while making the final product becomes more convenient for end-consumers (in particular the electrical voltage is increased, 0.7V is not very useful).

Requirements to meet these objectives:

- Scalability of production: economies of scale
- Sustainability: adequate supply of materials feedstock
- Ease of widespread use, market growth: ubiquitous deployment capability
- Standardization and quality control: reliability for consumers
- Safety

Module voltage ratings have increased from 600V to 1000V to 1500V, the commercial norm, to 2000V most recently. The higher the voltage, the less the current required for demanded power delivery – power losses scale by current squared.

Module Design and Materials

Early on, the PV module manufacturing industry standardized modules for its two main terrestrial markets of rooftops and ground-mounted project. Smaller customized modules are available for niche applications like portable gear but these products are a sideshow. The two dominant formats became 60-cell for rooftops and 72-cell for the ground (half-cut and triple-cut cell modules are variations of these two formats).

Larger modules are more economical for the manufacturer to produce (subject to controlling defects in the scaling up and the higher short-circuit currents from larger wafer areas), but bigger sizes become inconvenient for installers on rooftops. The smaller 60-cell module makes for easier handling by a single person, while the bigger 72-cell module requires additional labour or lifting equipment. The risk of incurring latent defects in a module during installation increases, with size and weight, in particular for rooftop work. Dropping a module or heavily leaning on it, awkward mechanical stress during its handling, may result in invisible micro-cracks that grow with time causing module underperformance or sudden breakdown. Another important reason for smaller sizes for rooftops is geometry: the smaller the module, the easier it is to fit more in limited space. Other than size, the 60-cell and 72-cell formats follow the same design pattern.

Modules consist of these basic components:

- The cells, patterned for the overall module layout
- Cell interconnections: soldered tab/ribbons, conductive adhesives
- Tabs (aka ribbons) to connect cells to current collectors
- Glass, for the front and possibly back as well
- Metal frame, if not a frameless glass-glass module
- Encapsulation: EVA, POE, TPE, other polymer types
- Backsheet
- Adhesives and edge sealants
- Junction boxes with bypass diodes, fuses, output cable
- MC4 external connectors

The module layout is rectangular in a typical portrait aspect ratio of ~2:1 (length to width for 72-cell, shorter length for 60-cell), with the cells connected in series in a string (aka branch) running along the length for the total module voltage, and the strings in parallel along the width for the total module current – e.g., six parallel branches of twelve cells each. A power cable for the total DC power output is housed in a junction box (or split junction boxes) at the back of the module, where the series and parallel connections come together, with a MC4 connector for the external circuit.

Glass

Most of a module is glass: ~65% by weight and ~85% by volume. Tens of millions of tonnes of PV glass are annually produced.[1,2] The dominant form of glass used for modules is tempered, rolled low-iron glass for greater than 92% transmittance, with an anti-reflection coating.

The chemical composition of industrial glass is mainly silica SiO_2 ~72-75%, soda ash Na_2O ~12-14%, lime CaO ~8-10%, metal oxide stabilizers alumina Al_2O_3, magnesia MgO ~2-5%, with trace impurities less than 1%. The trace quantity of ferric oxide Fe_2O_3 < 0.015% (< 15 parts per million), primarily distinguishes PV glass: a 10-20x lower content for high transmittance, low optical scattering, and spectral uniformity – avoiding the greenish tint of ordinary glass. Ideally, the silica is sourced from low iron mineral deposits (sand, quartz), or a costly purification step is required.

Rolling, rather than the float-glass process, enables tighter thickness control to +/- 0.1mm. The glass is textured for light trapping, with a 380nm wavelength cutoff protecting the module encapsulation from harmful radiation. The glass is then tempered, in annealing and cooling cycles, to increase its compression strength to greater than 100MPa: 3.2mm tempered glass is capable of withstanding hailstorms of 35mm diameter at impact velocities of 23-30m/s. The final step is a thin anti-reflection coating (e.g., porous SiO_2, or SiO_2/MgF_2) ~100-200nm, to maximise the optical length of the cell, achieving greater than 94%

effective transmittance, and glare reduction. The backside glass in bifacial modules is thinner ~2mm, omitting an anti-reflection coating to maximise rear-side light capture.

Titanium dioxide TiO_2 technology is beneficial. TiO_2-doped glass blocks and filters UV radiation below 380nm for added protection of the module encapsulation. A coating of transparent, photocatalytic sol-gel containing nanostructured TiO_2 can provide a self-cleaning capability against air pollution, breaking up organic pollutants that are washed away by rain.

Encapsulation

The encapsulation for the permanent bonding and sealing of solar cells in the module is vital for its durability. A polymer encapsulant covers the solar cells, cell interconnections, busbars, and cell-string connections. It is laminated between the front glass and cells, and the cells and backsheet on the rear (or cells and glass at the backside for bifacial modules). The encapsulant serves multiple functions of environmental and mechanical protection, including:

- Optical coupling for > 91% light transmittance in the 300nm-1200nm range
- Absorbs thermos-mechanical stresses: from thermal cycling (233K to 358K temperature range), hot and humid conditions, varying mechanical loads (5,400Pa dynamic and static loads), to prevent fatigue, micro-cracks, and other degradation mechanisms in cells

- Moisture ingress barrier, for < $5 g/m^2/day$ water vapour transmission rate (WVTR), to protect cell metallization from corrosion and prevent potential induced degradation (PID)
- UV stability: no yellowness from decomposition
- Strong glass/cell/backsheet adhesion > 30 N/cm peel strength (depends on polymer cross-linking, gel content > 60-75%)
- Dielectric insulation: volume resistivity > $10^9 M\Omega.cm$, prevents Potential Induced Degradation (PID)

Ethylene Vinyl Acetate (EVA) is most commonly used for the encapsulant, typically 0.4-0.5mm thick on either side of the cells – suited to fast lamination cycles: 12-20 minutes cycle at 423K. EVA can be shredded and reprocessed for recyclability. An innovation of the industry standard material called white EVA, loaded with titanium dioxide/barium sulphate fine particles ($TiO_2/BaSO_4$), increases rear-side reflectance in monofacial modules while in bifacial modules it is applied between the cell gaps to increase albedo light trapping. [3, 4]Another new market entry is an EVA-POE-EVA sandwich encapsulant.

Thermoplastic Elastomers (TPE)[5] of olefin block copolymers, e.g., ethylene octene, are being considered for EVA replacement. The trade-off is easier recyclability, lower water transmission rate, and no cross-linking for non-polar TPA (physically entangled polymer chains rather than chemical bonding), vs a lower cost for EVA (polar vinyl acetate group).

Polyolefin elastomer (POE) is chosen, instead of EVA, for HJT and TOPCon. POE has a higher volume resistivity, by an order of magnitude, to prevent PID in higher voltage cells. In addition, it is compatible with the passivation layers in TOPCon/HJT – no vulnerability to acetic acid formation from the acetate group in EVA. POE has a lower water transmission rate and PID resistance than TPA.

Summary comparison of the main material contenders (new hybrid materials and technical features are constantly being improved upon, with changing prices):

	EVA	POE	TPE
WVTR (g/m^2/day)	~ 2.5-4.5	~1-1.5	~4-8
Volume Resistivity (Mega Ohms * cm)	> 10^9	> 10^10	> 10^9
Lamination time minutes	~10-18	~14-22	~9-14
Recyclability	Possible	Possible	Easiest
Anti-PID	Moderate	High	Moderate
HJT/TOPCon compatibility	Risky	High	Moderate
Estimated Cost/Tonne	~ $600-$900	~ $1200-$1600	~ $1300-$1800

The Potential Inducted Degradation (PID) protection specifically refers to leakage currents (sodium ion migration from the glass to the cell), driven by a high voltage potential difference, reducing the solar cell shunt resistance.

Backsheet

The backsheet is applied at the back of the module for its outermost protection, for electrical insulation, environmental protection, and mechanical support.[6] Historically, backsheets have been a frequent cause of module failure. In glass-glass modules, glass serves as the backsheet. For other modules, the backsheet is a multi-layered laminate typically consisting of fluoropolymer coatings, polyvinyl fluoride (PVF), polyvinylidene fluoride (PVDF), for the outer layer, PET for the core, and EVA, polyethylene (PE), fluoropolymer on the inside to securely bond to the encapsulant. The backsheet provides protection of:

- Moisture barrier with water transmission rate of < $2 g/m^2/day$
- Dielectric strength > 15 kV/mm
- Peel strength > 30 N/cm
- UV stabilization withstands > 100 kWh/m^2 without cracking

For POE encapsulants, special primers are required for the backsheet bonding. Backsheets are thinner in the range of 0.2-0.35mm than glass (glass needs edge sealants), and reduce the weight of the module. Fluorine release from the backsheet can be used for UV-FL spectroscopy to monitor module degradation. Backsheet improvements are not only reducing cost, but rapid progress in environmental compatibility and even easier material recyclability easier.

Comparative overview of market priorities and technology trends:

	Fluoropolymer-based			Fluoropolymer-free	
	PVF	PVDF	CPC/FFC	PET	PO
WVTR (g/m^2/day)	< 1.5	< 1.8	< 2.5	< 3	< 2
Dielectric Strength (kV/mm)	> 15	> 15	> 15	> 15	> 15
UV Stability (kWh/m^2)	> 120	> 100	> 100	60-80	80-100
Peel Strength (N/cm)	> 30	> 30	> 30	> 30	> 30
Estimated Cost/m^2	\$2.5-\$3.5	\$1.8-\$2.5	\$1.5-\$2	\$1.2-\$1.8	\$1.3-\$1.8

Pros vs Cons:

PVDF-based: 30+ years durability, high UV, hot humidity tolerance vs higher cost and fluorine use (an environmental regulatory concern).

PVF-based: marginally more expensive than PVDF, higher WVTR vs higher mechanical and chemical stability, and less susceptibility to yellowing degradation.

CFC/FFC (fluoropolymer coated): no internal adhesive glue vs primer cost for encapsulant interface (CPC: polyester coating e.g. acrylic, TiO_2FFC: fluoride film).

PET-based: lower cost, easy recycling vs lower UV, moisture resistance

PO (polyolefin)-based: easy recycling, higher humidity resistance vs more complicated processing (co-extrusion of the PO layers).

The aforementioned encapsulation and backsheet materials are important contributions of the petrochemicals supply-chain to PV module manufacturing.

Adhesives and Edge Sealants

In addition to the EVA for bonding internal layers of the module during lamination, other adhesives, edge sealants are also required.[7] Silicone, epoxy adhesives for the junction box, with high volume resistivity matching that of the encapsulant, thermal (> 358K) and environmental tolerances, are crucial to avoid junction box delamination particularly in hot and humid operating conditions. Butyl rubber is applied around the module perimeter as supplemental environmental protection, particularly for moisture ingress protection. Silicone sealants are required for edge sealing in glass-glass modules: UV stable, with temperature tolerances of 233K to 473K. Polyurethane bonding is used for the module frame, capable of withstanding high shear strengths greater than 5MPa.

Frame

Module frames are typically made from anodized aluminium (15-20 microns oxide layer), though cheaper galvanized steel alloy is starting to replace it.[8] The anti-corrosive frame provides structural strength and reduces thermo-mechanical stresses on the module, with an easy fit into the racking structure. Glass-glass modules are available without metal frames, but are more vulnerable to damage during installation – require greater care in handling to avoid cracking. Modules with metal frames are less risky for installation. The metal is grounded by a lug or clip to the mounting frame (at the side or on the back - always check for it), to provide a safe low resistance path to earth for leakage currents. For glass-glass modules, the insulating glass prevents leakage currents but when it is wet there is a risk of a stray conductive path – a grounding lug or clip cannot be used on glass because of the localized mechanical stress.

Cell Interconnects

Cells are laid out in a module in series connected strings to add up the voltage, with parallel strings aggregating the current collection for the junction box. Silver fingers on cells, a grid of thin metal lines visible on the module, feed the current output into busbars. These busbars (BBs) are also silver conductors, though copper is being increasingly pursued for silver replacement. The BBs are typically 0.4-2mm wide (thickness 100-300 microns), with a resistivity of 1.5-3μΩ.cm (sheet resistance 0.05-0.3mΩ/sq), to balance maximum current collection with minimal optical losses from covering the module front. Multi-busbar (MBB) design has become the industry norm for modules: 9-16 busbars per cell (absent from the front of back-contact cells, and shingled modules discussed later have overlapping cells connected by conducting adhesives).[9] A greater number of thin BBs reduces electrical resistive losses without a net increase in optical losses (by busbars covering more of the module front). Shorter conductance paths from the fingers of a cell to its BBs, makes for lower resistive losses (fingers have a higher sheet resistance than the BBs). In addition, reducing the density and thickness of fingers offsets the shading from MBBs. Improvements in low temperature silver paste and screen printing were instrumental in MBB implementation. MBBs have boosted module efficiencies in the range of 0.5-1.5%.

Tabs, aka ribbons,[10] connect the front and rear sides of adjacent cells to each other - the positive and negative terminals. Soldered joints of the tabs link the BBs together on the series-connected strings. The front BB of a cell is connected to the rear contact of the next.

Tinned copper is used for the tabs: copper for the high conductivity, and tin coating for soldering compatibility, anti-corrosion protection. The tabs are narrow linear conductors, with flat, grooved, or cylindrical wire shapes. Flat ribbons typically have an electrical resistivity of ~1.7μΩ.cm, resistance per length of ~0.4-1mΩ/cm, are 1-6mm wide (depending on the required current capacity), thickness of 0.1-0.3mm, with a length matching the cell spacing and a 1-3cm overlap per cell connection. The cylindrical or round alternatives, diameters of 0.2-0.4mm, also consist of tinned copper, similar bulk resistivity as the flat options, offering lower resistance per length 0.3-0.6mΩ/cm and enhanced light trapping. Copper clad aluminium is being investigated as an alternative to tinned copper: lighter in weight, but has lower conductivity. Copper coated polymers for lighter, more flexible structures are also under consideration.

Junction Boxes

The junction box (JB) is the electrical interface of the module for its power output. JB inputs are electrical leads (conducting wires) from the series-connected cell strings or subsets of the strings, called sub-strings, with diodes in parallel for each of the inputs. These diodes are called bypass diodes.

If a cell in any string/sub-string of the module becomes defective or shaded, the entire string/sub-string changes from a power source into a damaging power sink. The shaded or defective area starts dissipating the energy from the rest, turning into a "hot spot". The hot spot can break the module glass, cause a short-circuit, and widespread damage. A bypass diode provides a lower resistance path - lower resistance than the shaded/defective cell(s) - to prevent the hot spot.

If all the string/sub-string cells are working properly, the bypass diode is in reverse-bias and inactive. When a string/sub-string is shaded or becomes defective, the voltage across the bypass diode switches polarity to apply a forward bias - automatically turns it on. Power flows through the bypass diode from the rest of the module, circumventing the shaded/defective portion.

Bypass diodes are rated at 2x the V_{oc} for reverse-bias, to avoid breakdowns, and 1.5x the I_{sc} of the corresponding sub-string/string. Diodes with metal-semiconductor junctions[11] have lower forward voltage drops ~0.3-0.5V and faster switching speeds for this function, than p-n junction diodes. The diode voltage drop has to be lower than that of the shaded/defective sub-string/string it is covering, but not too low to avoid being switched on by noise. Higher voltage drops result in greater heat dissipation and efficiency loss. Active diodes with forward voltage drops of 40-50mV are the state-of-the-art. A small control circuit monitors the sub-string/string voltage - when it turns negative, a MOSFET is switched on for the low-resistance bypass path. Active diodes increase the module efficiency in actual operation, with reduced heat dissipation, and are much more robust to noise. Module manufacturers market JBs integrating active bypass diodes as "smart junction boxes".

Most new modules have split JBs – the power output split across at least two or three JBs - for lower heat dissipation (from divided current output), and redundancy in case one of the boxes malfunctions. Split JBs suited the introduction of half-cut cell modules - dividing the total current among twice the number of strings in a module. Split JB design increases module durability, reliability, and power conversion efficiency.

The JBs contain fuses rated at 1.5x output current for safety, with $4mm^2$ copper cables for the power delivery.[12]

MC4 Connectors

The MC4 connector consists of a silver-plated "male" copper pin and typically a beryllium copper alloy for the "female" socket spring. The male pin has a crimped cable and locking clip for secure mating.

The female socket is spring loaded for a tight grip on the male pin, with silicone rubber O-rings for double-sealed protection against moisture ingress - small moisture ingress, during or after installation, can corrode the metal and crack the connector. Connector housing is made of polymeric material with strong anti-corrosion, UV, humidity, and thermal tolerances. MC4 ratings have increased to withstand a maximum continuous current of 55-60A, with a contact resistance < 0.5mΩ, a surge current of 100A for 1s, and system voltage of 1500V, meeting the requirements of higher wattage modules.[13]

CTM (cell-to-module) design innovation

A variety of design innovations have contributed to CTM efficiency gains, without disproportionately increasing the CTM production cost. Both optical and electrical losses have been reduced in modules – a concerted industrial effort over the past decade.[14,15] CTM improvements are a major reason for the dominance of mono c-Si technology. CTM is measured as the ratio of the combined power output of constituent cells to the aggregated power output of the individual cells:

CTM = the integrated cells module power/(number of cells • individual, standalone cell power).

A decade ago, the CTM for c-Si modules was ~88%. Modern module designs have pushed CTM to almost 100% and even higher for some designs and operating conditions. The emphasis is on CTM for actual operating conditions, rather than the STC.

Half-Cut Cell Modules

Half-cut cell modules are a resounding commercial success. Solar cells fit the size of the wafer, with ~0.5-1mm trimmed from the edges. Larger area wafers (M10:182mm x 182mm, M12: 210mm x 210mm) reduce the cell manufacturing cost, but also increase power losses to heat dissipation. Cutting the cell in half for the module is a clever way to take care of the problem. The current through a cell is proportional to its surface area. Resistive heat loss in the tabs connecting cells to busbars is directly proportional to the square of the current. Therefore, cutting the cell in half or by a third, reduces the tabs heat loss factor by a factor of 4 and 9 respectively.

Larger area wafers can be used for manufacturing cost reduction with less power lost to resistive heat dissipation, subject to two conditions: first, the larger area processing does not pick up defects, and second, the power extraction from the half-cut cells avoids the penalty of current collection from a larger number of cell strings (the number of cell strings will double by halving the cell). Laser cutting addresses the first condition: reduces cell edge damage and associated recombination losses. The cost of the second condition, increasing the number of tabs, is outweighed by the benefit of reduced resistive heat losses. MBB technology enables current halving in the tabs, by allowing for thinner tabs. The net result is a 5-8% module power output gain for 144/120 half-cut cell designs versus corresponding 72/60-cell versions. Another benefit is how the division of current collection enables split junction-boxes to reduce power losses from shading, individual cell defects, and improve redundancy. Half-cut designs, combined with MBB and split junction-boxes, strengthen module durability and reliability.

Unsurprisingly, most module manufacturers now offer half-cut cell modules. The market share of this module design in the space of a few years has risen to over 90% of total global sales volume.[16] The half-cut design can be taken a step further to triple-cut cells. It becomes technically feasible to push the CTM ratio above 100%. The trade-off is in the incremental production cost of the additional manufacturing complexity versus the gain in triple-cut power capacity. With improvements in laser cutting precision and tabs technology the market entry of more triple-cut modules is quite likely.

Shingled Modules

Another design strategy to raise CTM, approaching and above 100%, is shingled modules. Solar cells are cut into strips, 1/5th to 1/6th of the wafer area, 30-40mm wide, for overlapping placement of 1-2mm length (resembling shingles on a roof), connected in series by electrically conducting adhesives (ECA). No cell busbars or tabs are required. The entire module front surface is covered by photoactive layers of cells. Parallel currents of the series connected cells are aggregated by external wires at the back of the module to feed the junction-boxes. The ECA consists of epoxy resins, for the cell bonding, loaded with 80-90% silver flakes for high conductivity: resistivity of 5-10μΩ.cm (vs 1.5-3μΩ.cm Ag MBB), 0.25-1mΩ/sq sheet resistance (vs 0.05-0.3mΩ/sq Ag MBB), with the advantage of lower processing temperature of ~450K (vs 1000K for Ag MBB). The ECA has higher resistivity but shorter conduction paths resulting in lower resistive heating losses of ~0.5%, a power gain of ~1% over MBB modules of the same total area.

While most module manufacturers offer half-cut designs, a smaller subset produce shingled modules. Overlapping placement of the cells requires a challenging degree of manufacturing process precision. Shingled modules using HJT cells benefit from a manufacturing process synergy in low-temperature processing (ECA application temperature suits the amorphous silicon layers of HJT), and offer consumers a higher bifacial gain factor - shingled modules have a largely unobstructed rear side as well, maximising the HJT symmetrical passivation advantage for a 90-100% bifaciality factor. For monofacial designs, a combination of HJT and IBC cell technology for ECA panels is most advantageous. Shingled modules on the market are expensive: as much as 50% higher price per watt peak capacity to an equivalent half-cut panel.

The discarded silicon (kerf loss) from cutting strips is a small proportion of the silicon wafer cost - the main cost differential drivers are ECA materials and precision assembly. Shingled modules demand a premium for their sleek, all-black aesthetic appeal (5-10% thinner, 10-20% lighter than most of the competitive designs). This design is more competitive for the high-end rooftops market, and well suited to the new market growth of building-integrated-photovoltaics (BIPV).[17,18]

Interdigitated Back Contact (IBC)

Monofacial IBC modules, building on IBC cell technology for all back contact modules, are capable of over 100% CTM in particular operating conditions. A white backsheet provides an optical gain boost of 1-3%, for a higher monofacial CTM. IBC technology is versatile, capable of HJT, TOPCon, and PERC combinations for performance enhancement (as discussed in the previous chapter). Interdigitated p- and n-doped cell fingers (80-120μm wide) are connected to rear busbars (0.5mm-1.5mm wide, 10-20μm thick, with 1-2mm spacing) via solder pads (~1.2mm wide, 2.3mm long) running parallel to the fingers. The principal IBC advantage is being able to connect a greater number of full cells in series than half=cut cells of the same watt-peak capacity in parallel, within the same area, for a higher voltage and lower current. An IBC module may offer a V_{oc} higher by ~ 80%, lower I_{sc} by ~50%, than that possible for a half-cut module, or even shingled modules with parallel current branches for the same watt-peak power capacity. Higher voltage and lower current for the same power delivery not only reduce resistive thermal power losses in the module design but also enable low-voltage DC system-wide financial savings e.g. fewer strings for the given system capacity, use of higher power inverters with a reduced unit cost per watt, smaller gauge DC wiring, lower cable power transmission losses, and an extension of the system daily operating range. As module costs fall sharply with production economies of scale, the IBC design impact on overall PV system economics, scalability and flexibility, gives it a distinct competitive edge.[19]

Bifacial

Bifacial modules can achieve greater CTM ratios than monofacial, because of enhanced light trapping within the module, particularly from the rear. Superior CTM performance depends on various design and deployment factors e.g. an elevated system construction for the underlying surface albedo, cell spacing, materials like the white EVA for cell gaps, and the use of trackers. Hybrid module designs combine different methods of engineering electrical and optical gains, driving rapid progress in higher CTM ratios. The methods include anti-reflection, back reflection coatings for optical gains, novel cell interconnection architectures e.g. the use of ECA with full-sized cells in place of busbars (0BB technology) for electrical performance gains, and enhanced durability and reliability for lower annual power attenuation rates < 0.3% per year. Half-cut, shingled, and IBC module designs, the monofacial and bifacial versions, are all commercially available with 30-year bankable power performance guarantees.

CTM Assembly

The ratio of the total lifecycle cost of the module to its total lifecycle electricity generation is defined as its levelized cost of electricity generation: the module LCOE. Module energy yields have disproportionately increased, driven by higher cell efficiency and CTM innovation, compared to the corresponding cell and module production cost. As with the cells, the module BOM dominates total production cost. CTM assembly lines cost much less than cell manufacturing capacity.

The typical order of CTM assembly processes for a high-throughput factory, using parallel lines to accelerate production, is:

Cell inspection

- Flash I-V testing: 2,000 cells/hour, dark I-V (reverse-bias) testing: 1,800 cells/hour
- Optical imaging with automated AI image analysis algorithms for defects detection and characterisation (AOI)
- Electroluminescence imaging (EL), EL + AOI: 1,500 cells/hour

- Photoluminescence (PL) for recombination loss mechanisms: 200 cells/hour
- UV-fluorescence for anti-reflection coating, solder paste flux, broken cell fingers, surface contaminants: 1,000 cells/hour,
- Dark lock-in thermography (DLIT) for shunts – dark I-V + DLIT: 800 cells/hour
- Materials verification by random sampling: molecular vibrational spectroscopy (non-destructive) for doping profiles at 50 cells/hour
- Frequency domain infra-red scanning (FTIR) – destructive for passivation layers: 20 cells/hour
- Binning (classification) of cells by inspected I-V characteristics, quality grading

Cell Interconnections

- Half-cut cells: 1,800 cells/hour
- Laser scribing cells interconnected with copper tabs, soldering tin-Cu alloys at ~510-540K temperature for PERC, TOPCon and low temperature IR soldering of tin-bismuth alloys for HJT
- Soldering: IR or induction heating Pb-free solder
- IBC cells: 1,200 cells/hour
- Rear-side tabbing with ECAs for busbar free designs
- Precision alignment for pad to tab bonding
- Shingled cells: 900 cells/hour

- EVA bonding of overlapping cells
- Low temperature curing: 420-450K, 10-15 minutes for EVA, 410-440K, 10-15 minutes for POE
- 0BB: 2,000 cells/hour
- ECA dots in place of soldered busbars
- Electrical testing after interconnections:
- EL imaging to check for micro-cracks, e.g. from soldering, tab voids
- Series and shunt resistance measurements
- Robotic stringing: 1,200-2000 cells/hour
- Connecting tabbed cells in series
- Aligning cells, followed by soldering or ECA application
- Pre-stringing testing: I-V, EL, PL
- Post-stringing testing: EL, DLIT: 1,500 cells/hour

Layup of connected cells for encapsulation

- 1,500 modules per hour
- Glass loaded on to conveyor
- Front encapsulant application
- Alignment and placement of interconnected strings
- Rear encapsulant and backsheet application

Lamination: bonding of encapsulation

- 600-800 modules/hour
- Vacuum layering: remove air pockets, voids
- Thermal curing ~400-425K, 12-20 minutes
- Pressure application of 1-2 bar for adhesion

Framing and junction-box bonding

- 800-1,200 modules per hour
- Robotic clamping of anodized Al, metal frame
- Silicone for sealing
- Adhesive bonding of junction box
- Electrical testing: insulation resistance, wet leakage current

Final inspection and packaging

- EL imaging for defects: 1,200 modules/hour
- Flash I-V testing at STC: 1,200 modules/hour
- Steady-state illumination I-V under 1-1.5x Sun: 300 modules/hour
- Insulation test (1000V/1500V): 1,500 modules/hour
- Wet leakage current test: 600 modules/hour
- Random sampling for mechanical test:
- Peel strength testing: 1 out of 500 modules
- Damp heat cycling
- UV-B testing of encapsulation and backsheet durability (15 kWh/m^2, 48-72 hours, 333K)
- Different combinations of the cell inspection techniques can be applied to module samples

The rate limiting step in the assembly process is the encapsulation bonding. The lamination can be a single to six-step process, with or without the use of a membrane, depending on the laminator.

Glass-glass modules need extra care, for edge sealing, because the EVA cell encapsulant can permeate the glass and leak out.

For this CTM assembly example, the throughput is ~ 80 MW/week. The capex for HJT half-cut 80 MW/week manufacturing capacity would be around $25-35 million, depending on quality control measures. Investment barriers to market entry for module manufacturing capex are relatively low, but there is intense competition to face in succeeding with the investment.[20]

Production Cost Break-up, Module LCOE

Typical break-up of the total production cost for 80MW/week module assembly:

Module BOM: 85-90%

Equipment amortization: 4-5%

Labour: 3-5%

Energy: 2-3%

Overheads: 1-2%

Module BOM break-up range:

Cells: 50-55%

Cell interconnects: 4-5%

Glass: 8-10%

Backsheet: 4-5%, Glass in place of backsheet: 6-7%

Encapsulant: 8-10%

Frame: 5-7%

Junction box: 2-3%

Adhesives and edge sealants: 1-2%

The module LCOE calculation depends on the actual operating conditions simulated in a typical meteorological year (TMY). The module manufacturer provides a technical specifications file for insertion in an energy simulation software package to calculate its expected energy generation schedule. This information is used in conjunction with the module production cost to calculate the LCOE - the ratio of the module cost to its present value (PV) of energy generation:

Module LCOE = Module Production Cost/$\sum E_n/(1 + d)^{n}$.

Assume for a 700W_p module:

24.5% STC efficiency

0.25%/K temperature coefficient of power loss

1% first year degradation rate, linear 0.25%/year annual power attenuation for year 2-30

Global tilt irradiation of 1,650 kWh/m^2 per year

Annual average cell junction operating temperature of 313K (it will be higher than the ambient temperature)

Production cost of $0.08/W

For E_n the energy generation summation for each year, factoring in the annual power attenuation rate of 0.25% per annum (p.a.), we apply an arbitrary fixed discount rate of d = 5% p.a. (the discount rate reflects the time value of the energy i.e., greater kWh of energy in year 1 is worth more than that in year 30).

The formula gives module LCOE = $3.5/MWh, for our example.

In a later chapter, we shall go through the LCOE in more detail after accounting for total system cost and system energy losses. At this stage, the calculation is instructive to point out the module LCOE depends not only on the upfront cost per W_p, nameplate power or STC efficiency ratings, but the projected energy yield over its bankable lifetime. It is also important to note that the module energy generation covers the balance of system (BOS) cost i.e., the rest of the PV system apart from the module. The entire PV system performance crucially depends on the accuracy of the projected energy yield of the module.

The BOS budget is affected by module efficiency, e.g. a more efficient solar panel uses less space and reduces the cabling cost. The unit costs of BOS items are being continually reduced by technology improvements e.g., the materials for racking and control electronics. For the full system LCOE, the BOS and maintenance costs are added to the numerator. The LCOE formula is simply a geometric series calculation for projected energy production and costs. The calculation result depends on static assumptions e.g. a fixed discount rate for the time value of the energy generation, and a fixed annual power attenuation rate. The actual LCOE varies according to how true these static assumptions are in practice.

Module level power electronics (MLPE), i.e., sensors with control feedback, can adjust the power output for individual panel optimisation. This technology is commercially available but not widely deployed because of added cost and the masking of module defects. Module level sensing omitting the automatic power adjustment of MLPE facilitates string-level optimisation, by sharing module information for the string-level control. This option is also not common due to the cost of adding sensors to each module in the assembly process.

Projected Module Performance

The module LCOE calculation sensitively depends on its projected energy generation schedule, typically with a guaranteed linear annual power attenuation rate. The annual power attenuation rate has to be evaluated *a priori* for field performance in laboratory tests. There are two basic steps for this evaluation: identify and characterise all the possible degradation mechanisms for the module power output, then apply accelerated tests to quantitatively assess how the module responds to these degradation mechanisms. If the evaluation process omits a significant degradation mechanism, the projected power attenuation rate is suspect. The acid test is comparison to degradation mechanisms that actually occur in field-conditions. Field-testing is more important than what happens in tightly controlled lab conditions. Nevertheless, an initial lab evaluation is necessary to objectively estimate the module energy generation projection and hence its LCOE. Various test procedures are standardized for a like-to-like evaluation of projected performance.

Accelerated tests subject the module to cycles of severe operational conditions to assess its durability and reliability over time. These tests include combinations of thermal cycling: for temperature fluctuations over the guaranteed operating range, damp heating: high humidity and temperature, UV exposure, humidity freeze: humidity and thermal cycling, and high voltage stress. The module electrical parameters are measured for the various combinations. The emergence of any physical defects is monitored by imaging techniques that reveal minute details of the device e.g. electroluminescence (EL), photoluminescence (PL), and infrared (IR) – different types of techniques reveal different types of degradation mechanisms.

Accelerated test conditions are assigned a corresponding time period estimate: the specified lab testing conditions are equivalent to what the module might cumulatively experience e.g. over thirty (30) years. If the power output drops by 15% from its peak power output by a full set of combined, sequential accelerated tests designed for a 30-year specification, then the annual power attenuation rate is certified as 0.5% p.a. The accelerated test conditions are related to field-use by a theoretical model with some degree of experimental validation - requires the degradation mechanism to be modelled as a function of time. Test conditions are assigned to operational parameters so that the accelerated test time is quantitatively related to the expected module operational years.

For a well understood high-purity semiconductor like c-Si, thoroughly characterized industrial materials that have been used for years to work with it for processing, packaging, and protection - these accelerated testing models are more reliable. There is greater uncertainty over modules using materials that have a shorter history of industrial use in harsh, varying physical conditions.

Tandem Module Selling Price

We can work out an upper bound cost target or selling price estimate for a tandem module to compete with c-Si modules at the price of \$0.08/W_p, \$3.5/MWh LCOE. Think of the tandem top cell placed on top of the c-Si for the same BOS, but delivering an energy yield boost (similar to the DC: AC oversizing we shall later discuss, spreading additional energy yield over largely unchanged BOS cost). For simplicity, we assume the tandem module has the same power attenuation profile, bankable lifetime, temperature coefficient of power loss etc. (rather optimistic, best-case estimates) as the single c-Si module.

X = tandem system energy yield

Y = c-Si system energy yield

So that X/Y is the tandem energy yield gain

P = the tandem P-Si cost per W_p

Si = Si cost per W_p

T = total system cost per W_p

BOS = the rest of the system (except module) cost per W_p, with Si + BOS = T

The incremental tandem module cost divided by the incremental energy gain has to be at least equal to the ratio of the total system cost to energy yield for c-Si alone, or it won't be worth the addition.

To write it algebraically, $(P - Si)/(X - Y) \leq T/Y$

Re-arranging for P,

$P \leq T \cdot (X - Y)/Y + Si$ or

$P \leq Si + T \cdot (X/Y - 1)$

Since Si is a fraction of T, we can write $n \cdot Si = T$

$P \leq Si + n \cdot Si\ (X/Y - 1)$, or

$P \leq Si \cdot (1 + (X/Y - 1) \cdot n)$ where $n = T/Si = 1 + BOS/Si$

We see the lower the BOS/Si ratio, the tighter the upper limit on P.

Insert Si = $0.08/$W_p$, BOS = $0.16/$W_p$ for n =3 (module cost is around 30-50% the total system cost – a rough rule of thumb for typical system budgets). We take an energy yield gain of 1.4 (40%), equivalent to boosting single-junction c-Si 24.5% efficiency (commercial cells) to tandem 34.3% (small lab champion cells of P-Si), as proxy figures for actual energy yield gain. With our optimistic ceteris paribus assumptions, we obtain **P ≤ $0.18/$W_p$** (for n = 3).

This tandem module selling price target becomes tighter not only by falling prices of c-Si modules but also that of BOS items like inverters and mounting - all three are on a downward price trend for years. Prices for inverters and mounting, in particular ground-mounted trackers, have remarkably dropped. The trend is more than likely to continue with the expanding c-Si supply-chain economies of scale.

Quality Assurance

Modules are produced-to-order in batches. The fine print on a module datasheet indicates that the shown technical specifications might not exactly match any individual solar panel of the same model. It is therefore imperative that module purchasers know how to properly specify Factory Acceptance Tests (FAT) in comprehensive detail, including Acceptable Quality Levels (AQL) for different types of defects. At a minimum, the buyer must obtain flash test I-V reports and EL images for each module – the seller can easily provide this technical data as it is automatically generated during the production process. Buying modules without obtaining such basic information is naïve, and highly risky. An experienced buyer will specify detailed FAT and closely monitor its supervision. Every manufacturing batch has a certain number of defective modules that need to be identified and replaced before shipment. The FAT results, including BOM verification, are more important than checking certifications displayed on the module datasheet.

The modules that the manufacturer sends to internationally accredited testing laboratories, e.g. TUV/SGS, for IEC/UL certification might have an altered BOM or been fabricated with different machinery than the ones being purchased. It is up to the buyer, and their technical advisors, to ensure the modules are of proper quality meeting the advertised technical specifications.

Module warranties always include liability exclusions. The module manufacturer does not take liability for defects caused during installation, or due to mistakes in system design, the fault of other system components. Defects may occur during transportation or installation - the larger the project, the higher the occurrence of the latter. Latent manufacturing defects can have an adverse impact after a delayed period, secretly growing till the effects become apparent in the electrical performance data. Module batches can have systemic latent defects. The only way to distinguish latent manufacturing defects from other possible causes is thoroughly planned testing including Site Acceptance Tests on arrival of the goods, field testing during installation and post-commissioning for a Provisional Acceptance Test and Final Acceptance Test. Internal module defect degradation mechanisms are distinguished from external causes for warranty enforcement.

Modules are rated under standard testing conditions (STC) that are rarely replicated, if ever, in actual operating conditions - therefore the buyer, project developer, or installer must know how to calibrate the module performance from STC to actual operating conditions.

Unlike other power plants, medium to large-scale PV systems have thousands of energy generating units (the modules), not a single power conversion engine, to monitor. A defect that is not detected in time can become a degradation mechanism leading to other defects, making it difficult to identify the root cause. Defect effects may spread across different system components.

Types of Defects

Module defects have been extensively studied and classified, especially for c-Si modules. Module manufacturers maintain an extensive catalogue of defects for production quality control and deployment risk management.

With rapidly growing installations, longer deployment experience, different types of defects and degradation mechanisms are being understood in greater detail, so a knowledge of constant updates is essential. Defects can be peculiar to particular module designs and operating conditions. Lab testing has its limitations, for projecting the performance of new modules deployed in the field.

A non-exhaustive list of module defects to check for:[21,22, 23]

- Wafer defects: impurities, pinholes, dislocations
- Micro-cracks in cells
- Dead cells
- Shunted cells
- Cell mismatches
- Broken fingers

- Broken cell interconnects, tabs
- Short-circuited strings
- Busbar welding damage
- Soldering degradation
- Hot Spots
- Snail tracks
- PID
- LID, LeTID effects
- Glass breakage, corrosion
- Frame corrosion
- Encapsulation delamination, yellowing from photo-oxidation
- Backsheet degradation, cracking, discolouration
- Insulation resistance breakdown
- Faulty bypass diodes
- Junction box delamination
- MC4 cracking, loose connections

Early detection of the defect prevents it from becoming a degradation mechanism affecting other system parts. Identification and characterisation of defects enable effective warranty enforcement, and remedial measures to prevent its recurrence. Timely module replacement from on-site spare parts prevents system underperformance and loss of project revenue.

Field-Testing Techniques

No single diagnostic technique is capable of detecting and characterising all possible types of defects and degradation mechanisms. Reliance on electrical performance data by itself, is insufficient for leading indicators of growing problems. A combination of various techniques applied in a staged manner, in a comprehensive field-testing regime., with timely technical documentation of test results is required. We outline a number of field-testing techniques for modules, for more advanced analysis than the standard I-V data from PV module testers - methods capable of implementation by mobile testing rigs and/or drones:

Optical inspection, including artificial intelligence (AI) enabled image analysis

Visual inspection is a basic protocol, after unloading the pallets of solar panels (to check for damage during transportation) and the installation in mounting structures. For automated imaging, RGB digital cameras are typically used. These high-resolution, relatively low-cost cameras work in the 400-700 nm visibility range, and can deliver images with red, green, blue colouring. RGBD cameras add depth information to the image acquisition. RGB-IR cameras can use a colour array filter to simultaneously capture both visible light and IR.[24]

Method features:

- Operational real-time, non-destructive, non-invasive aerial scan

- Speedy data acquisition and analysis: one module per second

Application (types of defect detection):

- Video record of module installation, proper connections
- Identification of collected dust, soiling, bird droppings (persistent shading can lead to Hot Spots)
- Visual defects e.g. glass breakage, snail tracks (silver paste leak)
- Early sight of any mechanical deformations
- Signs of corrosion

Infra-Red (IR) Thermography

IR is a commonly applied field-testing method, with well-defined technical standards for its application.[25] Initially, handheld cameras were used for IR scanning. Drones with IR testing capability have become widely available.

Method features:

- Operational real-time, non-destructive, non-invasive aerial scan
- Data acquisition: one module per second
- Varied testing conditions e.g., at MPP, short/open-circuit, inverse current
- Rear imaging of module backsheet, junction box

Application:

- Whole cell: shunted, delaminated, partially shaded
- Part of cell: Hot Spot, crack, broken interconnect, soldering

- Substring: hot because of active bypass diode
- Pattern of hot cells: broken glass, short-circuited/shunted diode
- Potential Inducted Degradation (PID): hot close to frame

Lock-In Thermography (LIT)

LIT uses periodical pulsing of the heat in electronic devices for phase image analysis of the surface temperature variation. It is a mature technique for non-destructive testing of electronic devices.[26] Dark or illuminated LIT (ILIT) has been successfully demonstrated for in-depth analysis of PV module and cell performance during in-situ system operation.

Method features:

- Fast, non-destructive aerial scan
- Dynamic variant of IR
- No electrical contacts are required
- Module defects analysis in open-circuit conditions
- Thermal sensitivity of 100 micro-Kelvins

Application:

- Detailed analysis of solar cells within the modules e.g., shunt power losses
- Delamination of junction boxes

Electroluminescence (EL)

EL is similar to the PV process in reverse. A forward-bias electrical current is passed through the module to illuminate it, revealing defects.[27]

Method features:

- Rapid image acquisition is possible by a drone, but mobile testing rigs are more common as someone on the ground has to apply the forward-bias, set up the test conditions in any case
- Electrical contacts are required for the forward-biasing in test steps
- Performed at night for single modules
- Non-destructive technique
- Higher sensitivity to micro-cracks, sub-micron resolution from the ground
- Can also be applied to a whole string of the modules
- Enables detailed quantitative analysis of the properties of the modules and strings, with more accurate defects classification
- Application when the plant is not operational (offline)
- More time consuming than optical, IR inspections

Application:

- Cell cracks
- Dead cells
- PID (more accurate than IR for PID testing)
- Soldering degradation
- Silicon wafer defects
- Broken fingers, interconnects
- Busbar welding damage

- Failed diodes

Daytime EL is being developed, using a near-IR source to excite the modules.

Photoluminescence (PL)

PL imaging analyses the luminescence of the modules, when excited by a light source. Natural sunlight can be used to excite the modules during the power plant operation.[28] Images taken at different irradiation levels, during the power plant operation, are used to filter the luminescence for quantitative analysis.

Method features:

- Operational real-time, non-destructive, non-invasive aerial monitoring
- Highly spatially resolved information within strings, to cell-level
- Focus on cell performance within the module
- Analysis of series and shunt resistances
- No electrical contacts are required
- No external source is required for exciting the modules

Application:

- Cell cracks
- Light Induced Degradation (LID)
- Light and Temperature Induced Degradation (LeTID)
- Bypass diodes
- Junction boxes
- Electrical connectors

- Charge carrier induced degradation mechanisms

Ultra-Violet (UV) Fluorescence (FL)

UV-FL analyses the fluorescence from the fluorophores in module polymeric materials, when excited by a safe UV source.[29] Quenching and extraordinary brightening of the fluorescent signal indicates defects and anomalies.

Method features:

- Rapid monitoring, possible by drone technology
- Non-destructive technique, typically applied in dark conditions
- Adaptable to daytime conditions
- No electrical contacts are required

Application:

- Cell cracks
- Causes of the cell cracks, the history of defects in module materials
- Hot Spot identification
- Quality variations in the Bill Of Materials (BOM)

In addition, there standard electrical measurements[30] to check the health of the modules using a portable PV tester e.g. I-V curve traces (sunlit, dark), V_{oc}, I_{sc}, insulation resistance, and wet leakage current.

Field-testing is becoming increasingly important, rather than relying on lab testing and performance simulations. With new technology developments, the gap in accuracy and detail between field and lab testing is narrowing.

Most valuable field-testing techniques are non-destructive, non-invasive, and applicable in situ during the power plant operation. Granular, big data from in-situ field-testing, combined with increasingly sophisticated analytical algorithms, is able to deliver unprecedented levels of module performance feedback to manufacturers - a feedback loop for rapid design innovation and batch customization focused on actual operating conditions.

Professional Module Selection

The solar panel is the headline purchasing item for PV systems – the single most expensive item in the project budget. There is a lot more to its design, manufacturing, selection, and proper application than a superficial assessment. Label prices of cost per watt, STC efficiency, and power capacity headlines on datasheets can deceive - naïve buyers make module purchasing decisions based on these nominal captions. Projected module performance depends on the confirmation of defect-free products and analysis for the actual operating conditions.

Consider two modules X and Y, where Y has 23% STC efficiency and half the annual power attenuation rate of X e.g. 0.3% vs 0.6% for 30 years. The buyer bears a financial discount rate of 5% or 10%. Then module X must have an STC efficiency of 24.1% or 24.0% respectively to break even with module Y.

STC efficiency of module X = STC efficiency of module Y • [f(δ, n, d) for Y]/[f(δ, n, d) for X], where f(δ, n, d) is the present value of a function of δ, the respective linear annual power attenuation rate, "n" the number of years of the module deployment, and "d" a financial discount rate for the time value of money. This break-even condition for X based on STC rating and the datasheet attenuation rate is insufficient. Power loss due to temperature from STC is a substantial factor - calculated by simulation software from temperature dependence data provided by the manufacturer, so the comparative condition becomes:

Adjusted Energy Yield of X from simulation of actual conditions = (Adjusted Energy Yield of Y from simulation of actual conditions) • [f(δ, n, d) for Y]/[f(δ, n, d) for X], with temperature being the main adjustment factor from STC (irradiance intensity variation and spectral irradiance deviation are also simulated but have a lesser effect on the module power conversion efficiency).

In addition, the buyer has to check how the manufacturer's projected performance may vary because of actual operating conditions that standard simulations may not fully cover, e.g., transient shading from clouds or additional temperature rise in trackers (rises because of higher exposure to direct normal irradiance).

Professional module selection carefully assesses the multiple factors that we have discussed in this chapter e.g. the CTM design for actual operating conditions, type of bypass diodes, BOM verification, production line machinery (the laminator is especially important), the projected energy yield assumptions, ease of defect-free installation, and detailed FAT to field-testing validation. Module prices and purchasing contracts are negotiated accordingly. It shall become clearer in a later chapter why a simple calculation of the lowest simulated module LCOE by standard software is not necessarily an optimal condition. Additionally, optimal module selection requires system integration considerations, in particular the effects on LV system design. The I-V power control of the modules in their array strings by inverters is the next factor to consider.

--*--

Ch5 references, annotations, and supplementary reading

1. Major solar glass manufacturers, 80% of PV glass is produced in China:

   Xinyi Solar: www.xinyisolar.com - China

   Flat Glass Group: www.flatgroup.com - China

   Changzhou Almaden: http://en.almaden.cn/ - China

   Yao Jian Glass: www.yj-glass.com – China

   CSG Holding: https://www.csgholding.com/ -China

   AGC Solar: https://www.agc-solar.com/ - Japan

   Nippon Sheet Glass: https://www.nsg.com - Japan

   Saint-Gobain Solar: https://www.saint-gobain-glass.com/ - France

2. Review of issues and opportunities for glass supply for photovoltaic production at multi-terawatt (TW) scale, Sustainable Energy and Fuels, 2025, Tamal Chowdhury et al.

   DOI: 10.1039/d4se01567c

3. Hangzhou First Applied Materials, China, is the encapsulation market leader, especially in EVA and POE materials, by volume and tech innovation: https://www.firstpvm.com/en/

4. White EVA encapsulant:

   https://en.sveck.com.cn/products_detail/c_detailId%3D1061287734176055296.html

   Both Sveck and First Applied Materials, among others, offer the product

5. Mitsui Chemicals ASCE Solar product

   https://us.mitsuichemicals.com/service/product/solarasce/index.htm

6. Major backsheet manufacturers:

   Jolywood: http://www.jolywood.cn/ - China

Both fluoropolymer-based and fluoropolymer-free backsheets ~ 18% market share

Coveme: https://www.coveme.com/

Factories in Italy and China ~ 22% market share

DuPont Nemours: www.dupont.com

Used to be the market leader with its Tedlar product but not anymore ~ 3% market share

7. Henkel adhesives: www.henkel-adhesives.com - Germany

   Sika AG: https://industry.sika.com/en/home.html - Switzerland

8. Yinhe Solar: www.yinhealu.com – market leader in China, after Akcome collapsed for corporate restructuring (due to financial distress in other business lines)

   Mibet Energy: www.mbt-energy.com

9. SMBB extension of MBB technology: https://www.maysunsolar.com/blog-2024-guide-of-smbb-solar-cells-why-choose-smbb-solar-panels/

10. Ulbrich, light trapping solar ribbons: https://www.ulbrich.com/applications/fully-custom-solar-ribbon-for-special-applications/

    Sveck the encapsulant company is also a major ribbons manufacturer.

    Xian Telison New Materials is another solar ribbon specialist.

    Manufacturers are leaning towards copper coated polymers.

11. Metal-semiconductor junction diodes are known as Schottky diodes.

12. Specialist junction manufacturers include:

    QC Solar: www.qc-solar.com - China

    Amphenol Industrial Operations: https://amphenol-industrial.com - USA

13. Important, practical advice on MC4 connectors. A low-cost component but it can lead to major problems caused by poor crimping, dirty connections: https://rayenrg.com/common-faults-and-solutions-of-mc4-connectors/
14. Analysis of CTM electrical and optical gains: https://www.pv-tech.org/wp-content/uploads/legacy-publication-pdfs/1e4e2dd8c0-investigation-of-celltomodule-ctm-ratios-of-pv-modules-by-analysis-of-loss-and-gain-mechanisms.pdf
15. Fraunhofer ISE CTM design calculator, 2016: https://www.innovations-report.com/engineering/power-and-electrical-engineering/fraunhofer-ise-develops-smartcalc-ctm/
16. https://www.energytrend.com/news/20240312-45480.html
17. Analysis of Cell to Module Loss Factor for PV Shingled Module, New and Renewable Energy, 2020, Samchari Chowdhury et al. https://doi.org/10.7849/ksnre.2020.0009
18. CTM Power ratio of 97.7% for monofacial shingled PV module 100% employee-owned Chinese company, offering customized modules: https://couleenergy.com/unlocking-solar-efficiency-understanding-ctm-ratio-in-solar-panels/
    Shows CTM ratios have reached 104% for shingled modules
19. Study on the cell-to-module encapsulation losses of back-contact solar
    cell, modules, 2025, Journal of Renewable and Sustainable Energy, Yan Zuo
    DOI:10.1063/5.0246742
    Discusses metal wrap-through (MWT) and emitter wrap-through (EWT) back contact designs, alternatives to the IBC. EWT is an

extension of MWT. In EWT, a high density of small holes is drilled through the wafers, with doping inside to form an emitter connecting the front and rear of the cells. IBC is more advanced technology, with no vias or wrap-through contacts.

20. Turnkey PV module assembly lines are commercially available e.g.: https://www.mondragon-assembly.com/solar-automation-solutions/turnkey-lines-for-pv-module-manufacturing/
Jinchen Machinery and Zhengzhou Machinery are major suppliers of module assembly equipment, production lines in China
21. IEA Task 13 Reliability and Performance of PV Systems
ISBN 978-3-907281-71-0: Task 13 Report: PV Failure Fact Sheets (PVFS) Review 2025
22. Review of photovoltaic module degradation, field inspection techniques and techno-economic assessment, Renewable and Sustainable Energy Reviews, 2022, Lukas Koester and Sascha Lindig
DOI:10.1016/j.rser.2022.112616
23. A Review of Photovoltaic Module Failure and Degradation Mechanisms:
Causes and Detection Techniques, Solar, 2023, Hussain Al Mahdi et al.
https://doi.org/10.3390/solar4010003
24. A review of machine learning and drone-based solar panel inspection
techniques, IET Conference Proceedings, 2024, Madiah Omar et al.
https://doi.org/10.1049/icp.2025.0245
25. Infrared imaging of photovoltaic modules: a review of the state of the art and future challenges facing gigawatt photovoltaic power stations, Progress in Energy, 2022, Claudia Berhop, Ian Marius Peters et al.

DOI 10.1088/2516-1083/ac890b

26. Teledyne FLIR Systems is a commercial provider of tools for such testing:
https://www.flir.com/
27. Brightspot Automation is a specialist in mobile EL and PL field testing kits:
https://brightspotautomation.com/
28. Daylight PL imaging for module analysis:
https://userarea.eupvsec.org/proceedings/EU-PVSEC-2020/4CO.4.2/
29. UV Fluorescence imaging of PV modules, F. Vorster & Ernest van Dyck
https://sasec.org.za/papers2019/39.pdf
30. Basic PV module handheld testing kits are available from companies like Megger, HT Instruments, Seaward Solar, and Fluke

Ch6. Inverter

"Electronic circuit design is an art" – Paul Horowitz & Winfield Hill

The solar panel produces electrical power in a non-linear DC function of current versus voltage (modelled by the two-diode equation). Combined current and voltage output depend on sunlight and temperature, as well as module design characteristics, the physics of the device - a voltage limited direct current (DC) source. On the other hand, electric power grids worldwide have largely been built for AC power transmission and distribution. Precise frequency and voltage control are critical for the smooth functioning of the AC grid. An intermediate device is necessary to marry the electric power of the solar panel to the AC grid. This device is called an inverter: a circuit of power electronics, for the specific purpose of integrating PV modules with AC grids. It ensures optimal module performance during the DC to AC conversion for the grid. Quite understandably, it is referred to as the brain of the PV system.

The basic objectives an inverter has to serve:

- Track the operational point of the module on its I-V curve
- Control and maximise the power output of the module for the operational conditions (power maximisation is called maximum power point tracking: MPPT)
- Track the grid voltage and frequency
- Convert the DC power output of the module to AC power at the required grid voltage, synchronized to the grid

frequency with high power quality of the AC output waveform (no distortions in the sinusoidal V, I output)

- Minimise power loss in performing these functions

An inverter is the gatekeeper of electric power delivery from a solar PV array.

DC to AC

Sunlight is essentially like wireless AC power, sinusoidal oscillating E-B fields of electromagnetic radiative power, transferred in a mix of frequencies (photon energies) across empty space to our planet. A solar cell using a p-n junction converts the electromagnetic radiation to DC power output (though not like a p-n diode rectifier). This DC power has to be converted to AC power, in an electrical circuit, of a single fundamental frequency for the grid.

We shall illustrate for simplicity the DC/AC conversion process for single-phase flow. The conversion is done by what is called a full-bridge switching circuit, aka H-bridge circuit. Oscillating square pulses are generated from a DC voltage source (positive to negative voltage pulses of equal voltage to the voltage source: V_{in}). A pulse width modulator (PWM) generating a sine reference and triangular waveform carrier, modulates the bipolar square wave to approximate a sinusoidal waveform (the triangular carrier cuts pulses into sinusoidal form), followed by a low-pass filter to deliver a sinusoidal AC waveform of the target fundamental frequency.[1,2]

The PV module array acts as a current limited voltage source in this circuit, requiring a "boost converter" to transform its current output to that of a stable DC voltage source for the H-bridge circuit. The boost converter has an inductor in series with a diode, controlled by a switch (e.g. an IGBT or MOSFET), and a capacitor in parallel. When the switch is on, the inductor stores magnetic energy and when turned off, its stored magnetic energy is transferred to the electric field of a capacitor building up a higher voltage (in bursts of $V = L \cdot dI/dt$), than the input voltage V_{in}, across the capacitor. The diode blocks current from flowing back when the switch turns off. This boost converter is a standard circuit for stepping up the voltage of a power source, reducing current in the process, and turning the PV current-limited source into effectively a stable voltage source on the DC bus (sometimes called the DC link). The DC bus voltage is stabilized and stepped up in milliseconds, to ensure the inverter injects AC power at the required output voltage V_{out}.

Figure 7: simple H-bridge circuit topology with switches S1, S2, S3, S4 for single-phase DC/AC conversion (M is the load)

https://commons.wikimedia.org/wiki/File:H_bridge.svg

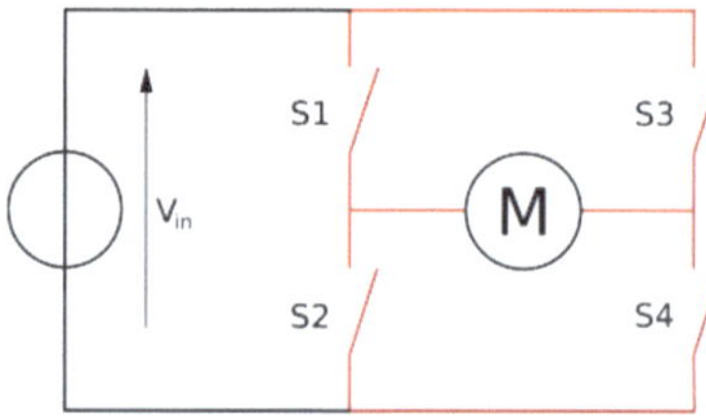

The load is connected as a bridge between the middle of S1, S2 and S3, S4. When S1 and S3 are switched on, S2 and S4 turn off, and the current flows across the load with positive voltage V_{in} (the source voltage of the PV module output from the boost converter). Then when S2, S4 are switched on, S1 and S3 turn off, the current flows across the load in the opposite direction with negative voltage V_{in}. An oscillating bipolar square waveform is set up. Capacitors are connected between the voltage source and ground, and across the switches for protection against voltage spikes and noise. Anti-parallel diodes are necessary across the switches to provide a bypass for inductive currents when the switches turn off. Insulated Gate Bipolar Transistors (IGBTs), with kHz frequencies, are used for the switches. IGBTs are preferred over unipolar MOSFETs for high-voltage and high-current applications (when the switching frequencies are below 20kHz) - due to their high-voltage tolerance and low forward voltage drop for high currents at high temperature. Enabled by industrial progress in silicon carbide (SiC) MOSFETs, inverter circuits have recently been improvised for higher efficiency, voltage, and power performance – multi-level hybrid topologies of IGBT and SiC MOSFET switches – e.g., in new 2000V string inverters delivering over 450kVA. The semiconductor layer between the MOSFET conduction channel and its drain is 10x thinner than that in a silicon device. Planar area of the conduction channel is expanded to increase the current carrying capacity - channel width is increased for short channel lengths. Multiple MOSFETs are connected in parallel to balance thermal effects for the delivery of high currents.

Next steps in DC/AC conversion are adjusting the frequency and shape of the oscillating bipolar wave of square pulses. Any periodic waveform is mathematically equivalent to a composition of sinusoids of different frequencies - square pulses are full of unwanted sinusoidal frequencies. The width of the pulses is controlled by the duration of the switching on/off time.

A duty cycle D is defined by $D = T_{on}/T$, where T_{on} is the time period the switch is on during the total period T of a frequency cycle, $T= 1/f$ (f being the frequency of the waveform).

T_{off}, the time period the switch is off = $(1 – D) \cdot T$.

$T_{on}/T_{off} = D/(1-D)$.

T_{pulse}, the width of the square pulse = $D \cdot T$.

For $T_{on} = T_{off}$, $D= ½$, $T_{pulse} = T/2$,

Or more generally for $T_{on}/T_{off} = n$, $D = n/(n + 1)$.

Modulation refers to the dynamic control of the time period pulse width and duty cycle. Control signals to the H-bridge switches are sent by a digital signal processor (DSP) in the inverter. The DSP is powered by less than 5W from the PV module array. It generates a reference sine waveform with a programmed voltage and target grid frequency (e.g., 50/60Hz), and a triangular waveform (carrier) of much greater frequency ~10kHz. Depending on whether the sine amplitude is greater or less than the carrier, control signals are sent to the switches to turn on and off - the switches need

Boolean digital signals to turn on/off. On and off signals start and end the pulse width duration: the longest T_{pulse} for sine curve peaks, shortest at zero crossover - at its peak, the sine wave stays above the carrier for the longest duration. The duty cycle is automatically adjusted by $D(t) = T_{pulse}(t)/T(t)$ at each instant, where T(t) is the carrier rate of generating pulses ~1/10kHz = 100μs. H-bridge output of varying pulse widths, and duty cycles, thus undergoes pulse width modulation (PWM) over a PWM cycle of 100μs. The average voltage, current, and power of the H-bridge output approximates that specified by the reference sine wave but with higher harmonics because of the square pulse shapes. Passed through a low-pass LC filter, the higher harmonics are filtered out to obtain pure sinusoidal voltage, current, and power output of a single frequency.

The LC filter cutoff frequency is set at ~1-2kHz for a grid frequency target of 50/60Hz. Most of the harmonics to filter are around the 10kHz frequency of the triangular carrier waveform generator and integer multiples. PWM does not generate significant energy in the frequency bands between 100Hz and 1kHz. A 1kHz cutoff has a negligible attenuation rate (~ 0.01dB) at 50/60Hz.

So far, we have been assuming the IGBT turns on and off instantly but it takes a finite time to switch. A small dead-time delay is intentionally designed in the switching between turning the switches on and off, to protect against the risk of S1, S2 or S3, S4 being on at the same time. Dead-time is kept to a minimum of 1-3μs, so that D(t) does not fall below a critical threshold in the PWM cycle. The average voltage in a PWM cycle depends on D(t). Average AC voltage over a PWM cycle = $V_{dc} \cdot (2 \cdot D(t) - 1)$ - the average output voltage of the H-bridge before filtering. The filtered AC voltage must always exceed the grid peak AC voltage = $2^{1/2}$ times $\cdot$ V_{rms} rating of the grid. If the voltage over a PWM cycle falls below the grid voltage, the power supply is interrupted.

We have described the DC/AC conversion for a single-phase inverter. A three-phase inverter DC/AC conversion operates on the same PWM principles but instead of a full H-bridge, it uses three half-bridges with two switches per circuit for reversing the current flow (a single DSP produces 120 degrees out-of-phase sine reference curves for each half-bridge, with a shared triangular curve generator: identical frequency, slope, and timestamp). The half-bridges are connected in parallel on the same DC bus to the PV source V_{dc}. Each circuit is driven by V_{dc} - the voltage is divided by two capacitors with two switches across the capacitors to generate square wave oscillations between $+V_{dc}/2$ and $-V_{dc}/2$. The half-bridges produce PWM modulated outputs that are 120 degrees separated in phase. Half-bridge output voltage is $V_{dc}/2$ with line-to-line peak voltage between any two of the three phases = $V_{dc}/2 - (-V\text{-}_{dc}/2) = V_{dc}$. The rms line-to-line voltage is given by $V_{rms} = m \cdot V_{dc}/2^{1/2}$, for a balanced load, where “m” is the modulation index between 0 and 1 (mostly close to 1). Low-pass LC filters the half-bridge output for the AC sinusoidal power flow at the target frequency.

Grid Frequency Lock-in

The frequency of the combined three-phase power output is tied to that of the grid by a phase locked loop (PLL) circuit in the inverter. Grid frequency fluctuates around its specified value of 50/60Hz - the inverter PLL is programmed to instantly adjust its frequency in sync with it. The PLL consists of a voltage measurement circuit, phase angle detector, loop filter, and a numerically controlled oscillator (NCO) driven by a DC input voltage signal. All modern inverters operate on digital PLLs i.e., digital signal processing for the control functions.[3]

The voltage waveform of the grid, or of the circuit connected to the inverter, is measured by the inverter: $V_{grid} \cdot \sin(2\pi f \cdot t + \varphi_{grid})$, where f is the frequency of the grid, φ_{grid} its phase angle, and V_{grid} the rms grid voltage. Capacitive dividers or high resistance voltage dividers step down the voltage. An isolation amplifier digitizes data streams across a high frequency capacitor or miniature voltage transformer (these devices provide galvanic isolation, blocking high frequency noise). An analog to digital converter (ADC) digitally formats the information for the microcomputer of a PLL digital signal processor (DSP).

The information is prepared for control signal input into the PLL DSP by standard mathematical methods for vector transformations. Three-phase voltage and current information is transformed to a stationary two-phase frame of reference: two vectors V_a and V_b of sinusoidal waveforms representing the same electrical information. Second-order generalized integrators, signal processing algorithms for frequency adaptive filters, correct for oscillatory signals from phase imbalances - grid harmonics are filtered to provide a robust reference frequency signal for the PLL. A further transformation is made to a rotating frame of reference, with a direct and rotating axis, at an angular velocity corresponding to the grid frequency. The second transformation gives V_d and V_q, where V_d represents the in-phase, steady-state DC voltage amplitude on the direct axis and V_q the quadrature voltage with phase angle information on the quadrature axis:[4]

$V_d(t) = V_a(t) \cos \theta(t) + V_b(t) \sin \theta(t).$

$V_q(t) = -V_a(t) \sin \theta(t) + V_b(t) \cos \theta(t).$

$\theta(t) = 2\pi f \cdot t + \varphi_{grid}(t).$

The phase angle error is $(\varphi_{grid}(t) - \varphi_{ref}(t))$, where $\varphi_{ref}(t)$ is the phase angle of the reference sine wave generated by the PLL. The phase angle error $\Delta\varphi$ is calculated by $\Delta\phi = \tan^{-1} (V_q/V_d)$. For small phase angle errors, it approximates to $\Delta\phi \sim V_q$.

The calculated $\Delta\varphi$ passes to a loop filter for error signal processing. A low-pass digital filter removes noise in the $\Delta\varphi$ signal, and a proportional-integral algorithm calculates a frequency correction from the $\Delta\varphi$ signal, for programmed conditions of the grid frequency, phase angle error margins, and response time. The corrected frequency information is sent to a NCO (NCOs replaced voltage controlled oscillators for this function) to generate a sine wave for the PWM circuit. Then PWM synchronizes power output to the grid frequency, making adjustments to frequency deviations within tens of milliseconds.

Single-phase inverters generate a virtual quadrature signal for V_d and V_q inputs into the PLL, based upon different signal processing methods, to follow the same frequency control process.

The PLL grid frequency lock-in process takes place in the timespan of 10-50ms, depending upon the harmonic distortions in the grid that have to be filtered and the magnitude of the phase angle error. For contrast, you will recall that the inertia times of synchronous generators driven by heat engines was on the order of a few seconds and their frequency response times ranged from 1-2s (OCGT) to several minutes (nuclear). The difference in respective speeds comes from the digital signal processing times of modern electronics and switching times of kHz IGBTs versus thermal and mechanical inertia.

Grid Voltage Tie

The inverter uses ADC data from the grid voltage measurement, to calculate a reference $I_{d,ref}$ and $I_{q,ref}$ in the quadrature rotating frame of reference: I_d and I_q are in-phase and out-of-phase current components respectively. Magnetic field current sensors or shunt resistors in the inverter measure its actual line currents: $I_{d,actual}$ and

$I_{q,actual}$. Differences between $I_{d,ref}$, $I_{q,ref}$ and $I_{d,actual}$, $I_{q,actual}$ are calculated by the proportional-integral method to send voltage adjustment commands of V_d, V_q (voltages in the rotating reference frame), their corresponding phase voltage values, to the PWM DC/AC circuit. The inverter voltage is instantly controlled for the grid voltage connection requirements.

The V_d, V_q commands to the sine reference curve in the PWM DC/AC circuit ensure that the AC voltage output is marginally higher than V_{grid}, the grid voltage, for uninterrupted power supply. This condition is ensured by a factor called the modulation index "m" - m ranges from 0 to 1 - the ratio of the peak voltage of the sine reference curve to the peak voltage of the triangular curve.

The modulation index tunes the amplitude of the sine reference curve for each phase so that

$V_{ref}(t)$ of sine reference curve = $m \cdot \sin(2\pi f \cdot t)$,

V_{peak}, AC phase = $m \cdot V_{dc}/2$ per half-bridge DC/AC circuit.

If the inverter voltage output is falling below V_{grid}, m is increased.

The modulation index is set by $m = 2 \cdot (V_d^2 + V_q^2)^{1/2}/V_{dc}$, where V_{dc} is the DC source voltage for the PWM DC/AC circuit.

The modulation index value is adjusted so that the phase voltages are always sufficiently high for active and reactive power delivery, without causing excess overvoltage. For a single-phase inverter, the modulation index "m" can be set directly without having to calculate V_d and V_q.

A more sophisticated voltage modulation method for 3-phase inverters is called the Space Vector Modulation (SVM) technique. Instead of tuning the voltage of each phase separately, SVM treats the three voltage phases as a single vector transformation in 2-d vector space. It synthesizes the optimal vector for all three phases, calculating V_d and V_q commands accordingly. SVM enables higher voltage utilization for each phase (the modulation index per phase can be increased to ~1.15), resulting in reduced harmonics and switching losses.

I_d and I_q are responsible for the active power P and reactive power Q delivery to the grid respectively - through V_d and V_q commands sent to the PWM DC/AC circuit. The inverter can rapidly control I_d, I_q, the corresponding values of line currents, for the grid voltage connection requirements set in the grid code. P and Q output are dynamically adjustable in 10-100ms. Inverter voltage and frequency are independently controlled for greater dynamic flexibility in responding to the grid. The inverter can adjust I_q to shift the current phase for Q absorption or injection. Automatic, dynamic P, Q responses are programmable for different conditions e.g., active power ramps for flexible dispatch, zero reverse power control, and reactive power injection/absorption for power factor or voltage level targets.

Maximum Power Point Tracking (MPPT)

MPPT is enforced in the boost converter, dynamically converting the PV current-limited source input into a stable voltage source (DC link) for the PWM DC/AC circuit. Voltage and current from the PV modules, inputs for the boost converter, are measured by voltage dividers and shunt resistors respectively. The voltage divider is protected from voltage surges by a transient voltage suppression diode. Magnetic field sensors are sometimes used for the current sensing (voltage induced by magnetic field of perpendicular current) in MW-rated central inverters, handling greater than 500A current magnitudes. The PV voltage and current measurements are sent to the ADC, for converting the measurement data to digital inputs. The DSP microcomputer estimates the operating point of the modules on the I-V curve, uses a MPPT algorithm to determine the operating voltage for the MPP, and sends a control signal to the boost converter to direct it towards the MPP. The boost converter adjusts its duty cycle to stabilize an input voltage to the PWM DC/AC circuit for the MPP. The standard MPPT algorithm technique is perturb-and-observe (P&O) – a simple, fast, trial and error method. Every 100μs -0.1ms, the boost converter duty cycle is moved up and down to see how it affects the calculated power output, with the duty cycle quickly adjusted for maximum power. The inverse of the inverter perturbation rate is sometimes referred to as the inverter frequency, not to be confused with its frequency rating of 50/60Hz for the grid. Faster perturbation rates can cause unstable oscillations around the MPP. Adaptive perturbation rates change according to the operating conditions: faster perturbation rates for volatile power, voltage, current variations, dP/dt, dV/dt, dE/dt (from rapid irradiance and temperature fluctuations), slowing down for more stable conditions. Calculations of dP/dV, dU/dV, guide the MPP determination:

At MPP, the slope of the P-V curve dP/dV = 0. To the left of MPP, dP/dV > 0, and to the right dP/dV < 0.

dI/dV = minus I/V. To the left of MPP, dI/dV > minus I/V, and to the right dI/dV < minus I/V.

Correlations between dP/dt, dV/dt, and dI/dt strengthen the analysis e.g., if both dP/dt and dV/dt are positive, it indicates the MPP is to the right. At MPP, the correlation will be zero.

P&O finds the local optima, with a risk of possibly drifting away from the global optimum of the I-V curve - particularly for sudden changes of the operating conditions e.g. shading from clouds or a module defect. Calculations of derivatives, and the rate of change of derivatives, helps distinguish between changes due to the perturbation method (being stuck on a local optimum) as opposed to PV operating conditions. There are different responses for a transient (cloud) versus persistent sudden change (e.g., a device defect). MPPT drift is curbed through this adaptable P&O.

An alternative MPPT strategy, to P&O, is to monitor voltage ripple introduced by the duty cycle of the DC/DC boost converter.

Instead of applying perturbations, these voltage ripples can be used in what is called Ripple Correlation Control (RCC) to estimate the proximity and direction of the MPP. Analog circuits can directly calculate correlations, while digital processing samples the phase differences between power, voltage, and current ripples for RCC. It is a speedier process than P&O, because no periodical perturbations have to be introduced.

Standard P&O might not work so well when modules in a string are shaded or develop defects, with the selected MPPT hiding the underlying problem. Many neural network algorithms, including particle swarm optimization approaches, are being continuously researched for superior MPPT under such conditions – the challenge faced by these algorithms is rapid convergence to the optimal MPP. There is a trade-off between speed and accuracy. P&O delivers an effective combination of easy implementation and rapid performance.[5]

Modern inverters are capable of performing I-V scans to diagnose PV modules, string-level performance. It takes only 5-10 seconds for the complete scan, so there is not much power interruption. The inverter does a sweep of duty cycles for the I-V curve scan, using the modulation index to stabilize the measurement conditions, V_{dc} in particular, making sure the scan does not affect the grid. The scan estimates V_{oc}, I_{sc}, and the shape of the I-V curve, identifying issues like anomalous series or shunt resistances.

Inverter Efficiency and Capacity

Inverters perform these functions with efficiencies of 96-99%, over a wide operating range of loads from about 20% of peak rating to their maximum capacity.[6] Power losses of 1-4% arise from switching and electrical resistance, increasing with higher temperature. At lower loads, less than 20% of peak rating, the fixed power consumption requirements of the control circuitry make the power losses and relative impact of switching losses more pronounced for a sharper efficiency drop.

The development of wide-bandgap (WBG) semiconductor technology, Silicon Carbide (SiC) and Gallium Nitride (GaN) devices, has delivered higher efficiency gains over 99%. Superior breakdown voltages, faster switching speeds, lower leakage currents than silicon, contribute to the greater efficiencies and power densities.

Inverter capacity is typically rated in total apparent power units of Volts-Amperes (VA), since it can provide both P and Q. Inverter capacities for P, Q are shown in a P-Q diagram of the operational space circumscribed by its total apparent power. The maximum limit of inverter capacity to provide reactive power Q, is the same as for active power P. **Inverters can operate anywhere on or inside the total AP capacity circle, with independent variation of P and Q inside the circle.**

Figure 8: P-Q diagram for an inverter specification[7]

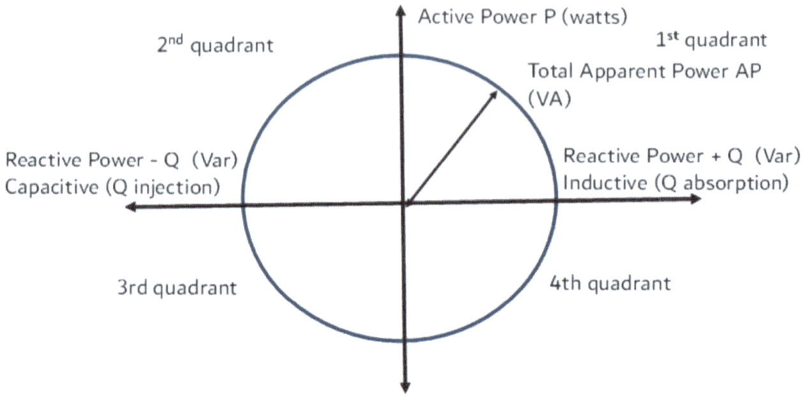

1st and 2nd quadrants: during the day, the inverter exports active power P from the PV modules to the grid/load, while supplying reactive power Q to maintain the necessary voltage level or other grid requirement.

3rd and 4th quadrants: at night, the inverter can also import a small amount of power from the grid to provide reactive power – known as the Q-at-night function. A modicum of grid power - less than 0.1% of kVA rated capacity ~0.5-1W/kVA - is drawn through a protected auxiliary circuit (not on the main DC to AC power path), enabling the inverter to act as a Static VAr generator (SVG) for its full Q capacity. Effectively, an advanced, faster version of a Static VAr compensator (STATCOM): <5ms vs 10-30ms response. "Static" does not mean that the VAr capability is fixed or static (like a passive capacitor bank), but that the device does not have any moving parts like a synchronous generator – on the contrary, the inverter is a highly dynamic and flexible Q instrument.

The voltage response of a synchronous generator is limited by the magnetization time constants of the rotor windings and the whole magnetic loop (the field between the rotor and the stator). PV inverters in SVG mode do not have any such handicap, and can therefore respond much faster with Q injection or absorption. Transistor switching speeds determine the response time.

The hardware ratings of the IGBT, MOSFETs, capacitors, and busbar current density, primarily determine the inverter operational capacity. Adequate cooling arrangements, fans and ventilation, have to be planned for the thermal management of the power conversion.

Inverter capacity is rated according to the ambient temperature range for its operational use. The three main categories, by inverter capacity and use-case:

Central Inverters:

High-power inverters for large blocks of PV arrays, with multiple module strings connected in series/parallel: 1MVA to >10MVA capacity per inverter. Exclusively used for utility-scale ground-mounted projects, connected to MV or HV grids. Modular central inverters, as an economical halfway house to string inverters, have been commercialised.

String Inverters:

Low to high power inverters for strings of modules connected in series for the inverter input: 1.5kVA-385kVA (the capacities are being continuously improved upon). Applications include utility-scale ground-mount, C&I ground/roofs, and domestic rooftop projects.

Microinverters:

Single inverter per module is the common use but some can operate dual or quad module inputs, depending on module size: 0.25kVA-2kVA. Mainly used for smaller rooftop systems, projects with severe space constraints.

Market leader:

String inverters dominate the market, for a number of reasons:

- Combination of economy, system flexibility and reliability
- More granular control of module performance than central inverters – multiple MPPT inputs per inverter, without the much higher cost per watt of microinverters
- High efficiencies similar to central inverters, but capable of more modular and scalable system fit
- Indoor or outdoor placement, close to module strings to reduce DC transmission power losses
- Lower cooling requirement than central inverters
- More ventilation for thermal management than microinverters

- Ease of operations and maintenance, easier to install and replace in a system with greater performance redundancy – if one inverter fails the whole system is not brought down

String inverter prices per watt sharply drop by capacity rating: e.g., a factor of ~18x from $0.35/W for2kVA to $0.02/W for 350kVA.

Safety and Reliability Features

Inverters play a crucial role in PV system safety and reliability:

Anti-islanding:

Prevents the system from feeding power into the grid during an outage. Detects excess voltage and frequency deviations, set by programmable thresholds, to shut itself off.

Fault ride-throughs:

The inverter is programmed so it does not turn off for limited voltage and frequency deviations: low-voltage, high-voltage, and low-frequency, high-frequency ride-throughs prescribed by the grid code. Wide-bandgap semiconductors enable more precise control for fault responses.

Leakage, ground fault, residual current detection:

Self-diagnostic checks are performed during startup e.g. for grid frequency synchronization, insulation resistance. When the inverter starts up in the morning, it detects leakage currents to raise alarms or shut off. Any leakage currents occurring later during the day, during periods of high irradiance, can be obscured (requires additional monitoring and testing). Measuring the impedance between the PV array and ground for ground fault currents on the DC and AC sides of the circuit. Detecting current imbalances with thresholds for alarms and tripping.

Electromagnetic shielding:

Ferrite cores on AC/DC cabling, wound chokes, metal covers for power circuits, protective layers to prevent electromagnetic interference is standard.

Overcurrent and overvoltage protection:

DC-side switches and fuses, a transient voltage suppression diode, built-in AC-side circuit-breakers, surge protection devices in the inverter provide protection from voltage and current spikes. The system is isolated from fault currents and voltage surges. The DC switch is turned off for maintenance routines.

Anti-PID (Potential Induced Degradation):

Potential induced degradation can occur because of voltage induced ion migration between the module and its supporting mounting structure, the ground. PID suppression technology is incorporated in the inverters, adjusting the voltages of electrodes in recovery cycles to eliminate ion migration, prevent module degradation.

Ventilation and fire safety:

Forced air convection by electrical fans, with thermal sensors for control feedback, is crucial for cooling string inverters. Central inverters may have liquid cooling systems. The inverter enclosure is designed for harsh environments, high tolerances to heat and humidity exposure. Alarms are triggered by signs of overheating.

Arc Fault Current Interruption (AFCI):

Inverters can detect DC arcs by a combination of magnetic coil sensing, fast frequency domain harmonics analysis, and artificial intelligence algorithms identifying abnormal current fluctuations, triggering alarms or tripping. DC arcs are particularly dangerous, since there is no zero-crossover point.

Real-time communication control and monitoring:

Inverters transmit real-time information at all times to indicate their operational status and that of the system. An internal hardware register keeps a log of the inverter performance, including any faults.

Standard information protocols are supported, e.g., Modbus TCP/IP, WiFi, and RS485 ports. Remote control of the inverters is available through a data logger, the inverter manufacturer typically provides, or Supervisory Control and Data Acquisition (SCADA) systems.

Careful attention to these technical features in the inverter selection, is of high importance. Selecting inverters on the basis of their capacity rating and price per watt only, is most inadvisable - a false economy and very risky.

Maintenance

Regular maintenance checks of inverters are crucial. Inverter user manuals have detailed instructions. Check for no ventilation blockages, dust accumulation, moisture ingress, and signs of damage due to weather events or small animal intrusion. IR scans to detect overheating.

Scanning acoustics imaging is a non-destructive technique for advanced diagnostics. Reflected ultrasonic waves with varying acoustic impedances provide information about the internal structure of opaque devices like enclosed inverters - can detect internal materials decomposition, capacitors condition, thermal cycling degradation of IGBT, soldering bond wire. Important insights to understand and monitor the more frequent failure modes in inverters e.g., electrolytic capacitors.

Inverters typically come with a minimum five-years warranty, but the warranty will exclude liability for defects caused by improper system design, installation, and operation. Inverter logs are downloadable to sift for a history of faults, the causes of malfunctions. Inverter manufacturers periodically update the inverter firmware (software) - it's important to check the latest firmware upgrades have been applied. Diligent SCADA or data logger monitoring for any inverter performance anomalies, particularly with respect to each other for the same operating conditions.

Hybrid Inverters

Hybrid inverters interface to both the grid and a battery energy storage system. The main additional hardware electronics component is a bidirectional DC/DC converter for power flow between PV-to-battery, and the battery-to-grid. The bidirectional DC/DC converter steps voltage up ("boost converter"), by reducing the current, and down ("buck converter") by increasing the current, from the PV DC link using synchronized switches instead of diodes to control the power flows.

A transceiver is added to the DSP microcomputer unit to monitor the state of health of the battery, for controlling the charging and discharging cycles. Sensors for the battery monitoring, e.g. for temperature, are additional hardware.

The control of the other inverter components is modified for hybrid application.

PWM DC/AC circuit, controlled for additional function of AC/DC rectification:

The half-bridge AC circuits (two switches for each phase) that convert DC power flows from PV-to-battery to battery-to-grid, can also work as AC/DC rectifiers - a separate rectifier and voltage down-converter circuit for the grid-to-battery power flow is unnecessary.

DC/AC grid-forming operational mode:

A digital signal controller synthesizes a high precision sine curve that is modulated with the high-frequency triangular curve generator to control the half-bridge switches, to deliver sinusoidal AC power from the DC bus input. The PLL block is not needed, in this mode. The LC low-pass filter removes harmonics. During a grid power outage, it sets the target frequency of the islanded electrical circuit.

The grid reference voltage is set by controlling the duty cycles of the DC/AC circuit. The capacitor bank on the DC bus plays a critical role in voltage stabilization, with its operation precisely controlled by the DSP. Capacitance is adjustable by switching the effective use of capacitors in and out of the bus bank.

Hybrid inverters integrated in energy storage systems (ESS) can also function for the main electric utility grid-forming and black-start capabilities.

DC bus upgrade:

The DC bus has to operate at different voltages, so the capacitance of the supporting capacitor banks may need an upgrade. Overcurrent and overvoltage protection is added, to protect against battery discharges. Forced air convection is reinforced, or heat sinks might be added, for additional cooling.

Control algorithms:

The MPPT algorithms are supplemented by algorithms for controlling the battery power flow, including bidirectionality. Battery cell and module level monitoring is incorporated.

Hybrid inverters cost 10-20% more than regular grid-connected inverters of equivalent power rating, from the same manufacturer.

With the sharp fall in battery prices, hybrid inverters are gaining market share - a higher proportion of the total string inverter sales volume.[8]

Technology progress in semiconductor power electronics is relentless, driving further price reductions, greater reliability, longevity, and sophisticated capabilities in PV inverters. Advances in tunneling magnetoresistance (TMR) for current sensors, and wide bandgap semiconductor devices for faster, more efficient switching devices are but two examples of it.

The conventional electric power establishment has consistently underestimated what can be achieved by the application of semiconductor electronics in this field. There are no moving parts in the electronics (apart from auxiliary cooling fans), subject to mechanical wear and tear, messy chemical reactions, or radioactive material concerns, and instead the autonomous electronic device performance is underpinned by fundamental principles of solid-state physics. Power electronics is ideally suited to the application of artificial intelligence (AI). Inverter performance and functionality shall definitely be further enhanced by AI - the fine features of inverter hardware design becoming more powerful and versatile by the application of AI optimization. Methods of large-scale electricity generation and electrical information processing are converging on the underlying foundation of solid-state semiconductor physics.

--*--

Ch6 references, annotations, and supplementary reading

1. Power Electronics: Converters, Applications, and Design

   Ned Mohan, Tore M. Undeland, William P. Robbins

   ISBN: 978-0-471-22693-2

2. H-bridge circuit: https://en.wikipedia.org/wiki/H-bridge

3. Phase-Locked Loops: Design, Simulation, and Applications, Roland Best

   ISBN: 9780071493758

4. Fundamentals of Power Electronics, Robert Erickson & D. Maksimovic

   ISBN: 978 3-030-43879-1

   The transformations are known as the Clarke and Park transforms, For the Park transform of V-d and V-q, think of the three-phase voltage as a single rotating vector with three components. The PLL function is to generate a d-q frame that rotates at the same angular velocity and hence frequency. When the two frames are exactly synchronized, the grid voltage vector has a constant value component on the direct d-axis and a zero-value component on the quadrature q-axis. Any deviation in frequency or phase generates a non-zero q-axis component as an error signal for the PLL correction.

5. A comprehensive review of maximum power point tracking algorithms for photovoltaic systems, Renewable & Sustainable Energy Reviews

   N. Kamarzaman & S. Mekhilef, DOI: 10.1016/j.rser.2014.05.045

6. Datasheets of major inverter manufacturers can be found at:

   Sungrow: https://en.sungrowpower.com/ - China

   Huawei: https://solar.huawei.com/en/ - China

   SMA: www.sma.de – Germany

   Fronius: www.fronius.com – Austria

Enphase: https://enphase.com/installers/microinverters – USA

Enphase specialises in microinverters

7. A Comprehensive P-Q Capability Study for Grid Interconnection of Inverter Based Resources Plan, Shahmur Rahman et al: https://par.nsf.gov/servlets/purl/10494707
8. Sungrow pitch for hybrid inverters: https://en.sungrowpower.com/blog-detail/5783/is-hybrid-inverter-better-than-normal-inverter-faqs-about-sungrow-hybrid-inverters

Ch7. Mounting Structure

"No one knows what he can do, till he tries" – Publilius Syrus

A PV system is not all about quantum mechanical semiconductor physics and clever electronic circuit design. There is one critical mechanical component of a PV system: the mounting structures that hold the solar panels together in a stable array through varying weather conditions, enduring extreme weather events, for the project lifetime. Mounting structure designs are broadly divided for the two main PV markets: ground-mounted and rooftop projects. A third fast growing market segment is floating solar power on water surfaces, with the solar panels mounted in structures called floaters.

The mechanical support structure for the solar panel is perhaps the most neglected technical element in a PV system. More attention is paid to the solar panel and inverter, in the naïve belief that the mounting structure is simply like a piece of furniture to tightly hold the solar panel. In practice, it is much easier to replace solar panels or inverters than the mounting structure. Of all the major components of a PV system, a longer lifetime of the mounting structure without any repairs or replacements is most convenient.

The consequences of amateur or negligent mounting design can be devastating. Apart from compromising the overall stability of the PV system, it can result in accidents of bodily harm.

Ground-Mounted Systems

The mounting structure design has to satisfy a number of objectives:

- Securely holding the solar panel in a uniform load distribution, without any residual thermomechanical stresses on it[1]
- Protecting it from varying static and dynamic mechanical loads
- Bearing the static load of the panel and any snow, dust
- Bearing the dynamic load of wind, rainstorms, dust storms
- Anti-corrosion, high tolerance to heat, humidity, UV radiation
- Enables solar panels to be densely packed, easily connected electrically, with optimal orientation to the Sun
- Speedy installation, by standard machinery, over wide areas
- Adaptability to different topographies
- Allows convenient operations and maintenance (OM), access to the solar panels, cabling, for regular checks and corrective actions in case of any problems
- Compatibility with soil conditions
- Sustainable and widely available materials for scalability

The relative position of the Sun to any point on Earth, the angle of incidence of the solar rays, changes during the day, and from one day to the next over the course of the year. The orientation of the mounting structure towards the Sun is set by two angles: 1) the tilt angle, i.e. the vertical angle between the plane of the solar panel slot and horizontal ground, and 2) the azimuthal angle, i.e. the horizontal angle between the projection of a normal vector from the solar panel plane to the ground and the geographic true north of the globe (not the magnetic north). For any given location, these two angles specify an optimum orientation for maximum photovoltaic energy yield from a solar panel over the course of a calendar year. Assuming there are no shading or land topography constraints, the optimum azimuthal angle for a single solar panel in the northern hemisphere is 180 degrees (true South) and 0 degrees (true North) for the southern hemisphere. The optimum tilt angle for a single solar panel depends on the exact longitude and latitude. A rule of thumb for the optimum tilt angle of a single solar panel is the location latitude, with a seasonal adjustment of plus 10-15 degrees for the lower winter Sun and minus 10-15 degrees for the higher summer Sun.

The solar market grew by planning fixed-tilt structures for solar panels. A single tilt angle to maximise annual energy yield. The PV mounting array has to optimise the tilt angle not just for a single solar panel but all the panels comprising the array. Solar panels installed in a row shall face a certain limit, either by the available space or before the mechanical structure becomes unwieldy, so multiple rows have to be planned. If the rows are placed too close together, the front row shall shade the second row, and so on. An inter-row spacing is needed to prevent shading. Inter-row spacing is also needed to a certain extent for maintenance. A ground cover ratio (GCR) is defined as the ratio of the total area covered by the normal projection of the solar panels to ground, to the total area of the system installation site. GCR depends on the tilt angle, how many panels are installed in the racking of a single row (whether in portrait or landscape format), and the topographical constraints of the land. Solar array layouts with different tilt angles, GCR values, can be assessed for the given land. Energy simulation software, for the given site coordinates, calculates an optimal tilt angle for a single panel – but that angle might not be optimal for the whole solar array. Optimal tilt and azimuthal angles for the whole array maximise the energy yield from the total available site area, not just for one solar panel located in it. The array might consist of sub-arrays of varied tilt and azimuthal angles, adapting to land obstacles and the topography. For a naturally sloping area, like a hill, the optimal tilt angle may adjust to its slope. The overarching rule is minimisation of the LCOE ratio: minimising the total cost numerator, including array installation cost, and maximising the total energy denominator, strongly dependent on array orientations. Solar arrays with a fixed tilt and azimuthal angle are called fixed-tilt structures.

The fixed-tilt structure design will consist of a racking of solar panels in a row, normally one to two, sometimes three, denoted 1 x N, 2 x N where N is the number of panels in the row. For panels placed in the portrait format, a 1P or 2P abbreviation is used. More than three panels in a column are not recommended, because the structures become more unstable with height (distance from the foundation), with greater difficulty in secure installation and maintenance.

The standard material for mounting structures is galvanized steel alloy. Steel is a broad term for a class of metal alloys. The exact composition of the steel alloy makes a big difference. So does the galvanization method and coating thickness. Mounting manufacturers have developed proprietary formulas for longer lasting steel alloys and its galvanization.

The basic inputs required for the mounting structure design:

- Solar panel physical dimensions, weight, material, frame
- Dynamic and static load tolerances: typically set by wind speed and snow
- Soil conditions, obtained from geotechnical tests, geological data
- Site topography, land demarcation, location of the grid connection point
- Seismic data, flooding risk (1 in 100 years)
- Initial overview of a solar array layout

The mounting manufacturer can then design a fixed-tilt solar array structure and spatial layout for the PV system.

Foundations

Mounting designers have a dedicated engineer to calculate the foundations required for the fixed tilt structures.[2] The three most common types of foundations are pile foundations, ground-screws, and ballasts. The purpose is transferring mechanical loads on the solar structure, to the foundation to withstand. Each leg of the mounting structure fits in the foundation.

Pile foundations:

Pile foundations typically consist of cylindrical mortar, surrounded by a rebar steel wire cage, placed in the ground for in-situ hardening. Steel piles can be directly embedded with a cap, but cast-in-place concrete mortar around it adds lateral stability. The required diameter (~100-300mm) and depth (~1-3m) combination of the pile foundation is calculated from the results of standard push-pull tests (include bending, torsion), for the given static and dynamic load tolerances. Pile foundations are labour and machinery intensive, relatively time consuming (the time taken for mortar to cure), but work in just about any terrain. Mortar with high compressive strength and material stability is required e.g., C25/C30 grade concrete or higher. The use of pile foundations is a standard practice in the construction of buildings, so it is easy to source the skilled labour and materials for it.

Ground-screws:

A ground-screw consists of welded helical plates on a hot-dip galvanized steel shaft ~ 1-3m long, 50-100mm diameter, 10-

20mm thick plates, 50-100mm pitch for multiple spirals. The ground-screw is ideal for soil that is not too hard, but rough enough for traction by the screw. A pneumatic hydraulic drive can install ground-screws in a faster workflow than pile foundations, with less labour involved. No time is needed for mortar curing, though in certain soils, mortar might be poured in to cure around the ground-screw to securely fix it in place.[3]

The penetration depths of pile foundations and ground-screws can be varied in a solar array for alignment – very useful for sharper slopes and uneven topographies.

Ballasts:

Unlike the pile foundation or ground-screw, a ballast is placed above the ground. The ballast could be a concrete block, or open tray filled with gravel, heavy compacted aggregates of dirt, with attached weights if necessary. There is no ground penetration so it is easier to clean up at the end of a project, but it is typically more expensive and difficult to align on uneven land. For certain sites such as landfills, where ground penetration is hazardous or forbidden, ballasts are the only possible choice.[4]

Ballasts and ground-screws have less of an environmental impact than pile foundations.

Design and BOM Quality Control

The mounting structure on top consists of a long list of precisely machined and tested parts, e.g., including:

- Impact embedded pile
- Front and rear columns (vertical structural supports)
- Purlins (horizontal structural elements), purlin support, connectors
- Straight or oblique beams
- Pull rod (secures tilt angle), pull rod connector
- Braces, brace connectors
- Hex bolts and nuts, socket screws
- Flat and spring washers
- Clamps
- Conductive thorns for grounding

There are detailed material specifications, sizes, gauges, weights, for each part, mostly hot-dip galvanized steel, the bolts, washers, and nuts made of carbon steel. A weak link can undermine the mechanical integrity of the whole structure. Amateur or underqualified mounting builders might add bulk, to make up for lack of careful design calculations, but that can make the design worse. Structural analysis factors in data inputs of the mounting design, spacing of the structural elements, pressure components, wind directionality, wind speed, 3s gusts, displacements, for the maximum compression, lateral, lift loads, and maximum moment. If the mounting buyer does not see this basic detail, they should be worried.

As for the modules, the FAT results are more important than third-party certification shown for the general designs. The FAT ought to cover materials verification (alloy composition, galvanization thickness, bolts and nuts classes), the structural analysis calculations (match the manufactured product physical dimensions), oversight of tests, review of characterisation evidence for:

- Elastic modulus
- Shear modulus
- Elongation at breaking point
- Fatigue strength
- Shear strength
- Tensile strength
- Ratio of lateral strain to axial strain
- Specific heat capacity
- Thermal conductivity
- Melting solidus
- Electrical conductivity
- Anti-corrosion, with reference to corrosion class

Detail of the structural calculations:

- Static load calculation with reference to length, weight, and quantity of components
- Dynamic load calculation with reference to length, weight, and quantity of components
- Seismic load calculation
- Strength calculation
- Horizontal and vertical support forces

- Bending stress and moments
- Buckling stress
- Structural analysis simulations by industry standard FEM software
- Load bearing stress test, including inhomogeneous mechanical loading, push and pull forces
- Compression test (dislocation and maximum deformation forces)
- Results of the calculations aligned to the building code of the site jurisdiction

Fixed-tilt structures have progressed through several innovations e.g., the design of structures that can be dismantled (apart from the foundation) for re-use, and flexible connecting cables for installations on very steep slopes. Elevated canopies are routine now, with some designs enabling adjustable tilt of the shed.

CFD and Wind Tunnel Testing

For the risk management of large projects and top tier mounting design, greater sophistication is warranted.[5,6,7] Building codes specify wind speeds, attack angles, but it makes for static or quasi-static calculations, e.g., using static pressure coefficients, that do not account for the complexity of fluid dynamics, the damage complex fluid flow might cause in solar arrays. FEM models are insufficient for analysis of transient wind attacks, vortex shedding, edge effects, coupling between the dynamic wind distribution and the solar array structure for destructive effects, and cumulative fatigue. Computational Fluid Dynamics (CFD) simulations are used, by more advanced mounting designers, to investigate such aerodynamic issues. Pressure distribution maps, turbulent flow effects, and the risk of higher lift forces at the array edges are more informatively calculated. CFD, like all numerical simulations of complex physical phenomena, has its limitations. Aeroelasticity, mesh assumptions, boundary layer conditions, among other factors, pose difficulties.

Wind tunnel testing is the gold standard for quality control of mounting design, risk management of projects with very large solar arrays. Static and dynamic instabilities can be investigated. Aeroelastic models are carefully constructed, replicating the distribution of masses, moment of inertia, and stiffness of the array design, for a natural frequency much higher than the natural frequencies of aerodynamic interest. Instabilities such as vortex induced vibrations, resonant vibrations between the natural frequency of the array and vortex shedding frequencies at different wind attack angles, come under scrutiny. Another example is torsional flutter that can occur when the panels elastically twist under torsion, extracting energy from the wind flow to cause damaging vibrations. These instabilities are possible when the oscillating wake from one row of panels impacts another. The effects of integrated static and dynamic loading on the array, taking wind acceleration into account from the geographical terrain or built environment, can be probed more effectively. Wind tunnel testing is particularly valuable for optimising the deployment of tracking mounting structures i.e. trackers.

These instabilities, vibrations, may not collapse the array, but could cause micro-cracks or other defects in the modules, resulting in chronic system underperformance and faster system degradation. Unexpected vibrations in solar arrays is a crippling problem, that is very difficult to solve, requiring an expensive overhaul of the system.

Trackers

Mounting structures that can track the daily movement of the Sun have become the preferred option for ground-mounted projects, where the terrain and topography permit it (typically the steepness of the N-S slope).

Trackers have a greater impact in areas with high Direct Normal Irradiance (DNI), especially with bifacial solar panels (an additional 5-10% energy boost per year).

A variety of trackers includes:

Single-axis horizontal trackers, with east-west movement around a rotation axis parallel to the ground:

- Azimuthal tracking of the Sun
- 15-25% higher annual energy yield than fixed-tilt for the same watt-peak capacity of modules
- Lower cost, easier to install and maintain than dual-axis
- Greatest value-add in latitudes below 40 degrees N or S

Single-axis tilted, with a rotation angle tilted 20-30 degrees from the horizontal:

- Hybrid of east-west movement
- Suited to high latitudes above 40 degrees N or S, for geographies with low angles of incident sunlight
- 15-25% more annual energy yield than fixed-tilt, for the same watt-peak capacity

Single-axis vertical, with a rotation axis perpendicular to the ground:

- North-south movement around the vertical axis
- 8-12% higher annual energy yield than fixed-tilt, same watt-peak capacity

Dual-axis rotate for both tilt and azimuthal angles to exactly follow the Sun all year round:

- 30-45% higher energy yield
- But the most expensive

Single-axis horizontal with backtracking, is the most commercially attractive choice at present:

- Backtracking: panels tilt westward at dawn, eastward at dusk, to minimise inter-row shading, returning to optimal perpendicular alignment at peak Sun
- Superior combination of high GCR and high energy yield

Back-tracking single-axis horizontal trackers are the most compact for higher energy yield from limited space. The trackers are designed for trivial energy consumption, compared to the added energy production. AI algorithms enable enhanced tracking adjustments for different weather conditions, and safety response to extreme winds. The backtracking single-axis horizontal trackers can support 1P and 2P mounting. Single-axis horizonal trackers are available for an incremental cost of only \$0.01-0.02/W_p more than fixed-tilt. In return, the incremental energy gain is 15-25% higher with monofacial solar panels, 20-35% with bifacial, varying by the site location, terrain, and topography. HJT bifacial solar panels are ideally suited to trackers because of their higher bifacial gain factor, and lower power loss with temperature of 0.25%/K (greater exposure to direct normal irradiance, DNI, increases the average solar cell junction temperature).[8]

Rooftops

Anodized aluminium is the main material used for fixed-tilt rooftop mounting (trackers are not suitable for rooftops, though some designs allow an adjustable tilt for a seasonal energy gain ~ 4-5%). The exact alloy composition and thermal treatment is important, mostly from the 6xxx series containing magnesium and silicon e.g., **6005-T5/T6:** most commonly used for structural elements of rails, beams, and purlins, **6061 -T6:** higher tensile strength of larger structures, **6063-T5/T6:** preferred for complex shapes, components like lugs or clamps, aesthetics of lighter and sleeker designs, and **T66:** for more challenging conditions of heavy loads on tall buildings. Each variant has to be assessed for a target set of mechanical properties. Anodization of a thicker oxide layer, 10 to 25 microns, reinforces surface hardness and anti-corrosion.

Recently, fiberglass composites have been introduced by major manufacturers for mounting parts, with the principal advantages being lighter weight and lower cost.[9]

All of the quality control requirements spelled out for ground-mount, stringently apply to rooftop structures as well, with additional care in the structural calculations for the height. CFD simulation and wind tunnel testing are even more important because of the roof height. More serious accidents can occur from flawed rooftop mounting designs, particularly from panels lift-off. Ballasts are more frequently deployed for rooftops. Screws are often used for concrete flat roofs, with careful attention to the risk of water leakage. Rooftop mounting manufacturers pay much attention to the ease, speed of assembly and installation, e.g., the use of only one or two tools required for installing the mounting.

There is an increasing variety of rooftop mounting designs, with standardized versions for different types of roofs e.g.,

Hangars, typical on C&I, industrial buildings, warehouses:

- Use of heavy aluminium rails, typically spaced 1-2m apart
- Clamps or stainless-steel bolts for fasteners
- Rails bolted to purlins
- Corrugated roofs: rails on sheet with side-clamps, trapezoidal adapters
- Cheapest type of mounting ~\$0.02-\$0.03/W_p

Flat roofs:

- Triangular brackets with modules facing in an E-W direction
- Ballasted or anchored by screws
- Low tilt angles of 5-15 degrees, adjustable tilt is possible
- Enables denser packing of the modules, higher energy yield per unit area
- Can saves on inverter cost, MPPT sharing is enabled by E-W orientation
- Cost ~\$0.03-\$0.05/W_p
- Higher tilt 10-30 degrees A-frames are also used in an E-W orientation, but can take up twice the space and are more expensive ~\$0.05-\$0.07/W_p

  More labour intensive and time consuming to install

Sloped roofs:

- Tile hooks: curved metal brackets that latch under roof tiles
- Waterproof metal strips under the hooks to prevent water leaks
- Rails mounted on the hooks
- L-feet (angled brackets) bolt rails to rafters, with waterproofing seals – used for shingle, tile roofs
- Modules clamped to the rails
- Rail-less designs: hooks fixed to module frames, cheaper than rail systems

- Cost ~\$0.05-\$0.08/W_p

Elevated canopies of solar panels on roofs are deployable, like a carport but on the roof, subject to structural reinforcements. Raised ballasts for roof gardens is another design. Structural assessment of the roof load-bearing tolerance is a preliminary step for planning all types of designs.

Floating Structures

Floating PV projects have proliferated in recent years, not only on inland water reservoirs but offshore, with increasing 100x MW_p capacities.[10,11,12,13]

A breakthrough was the design of anti-corrosive, lightweight, high-density polyethylene (HDPE) floaters, costing \$0.06-\$0.08/W_p - the market leading variety as of now. Other designs consist of steel and aluminium structures for large offshore platforms, elastic membranes also for offshore deployment, and fibre reinforced polymers for HDPE replacement in extreme environments.

The floating structures are available in both N-S and E-W orientations. Feasible tilt angles are lower than that for ground-mount. Low tilt angles are preferred for stability against heavier waves and winds. Structural elements absorb mechanical stresses from water motion, so that solar panels are not damaged. Modular HDPE~ 5-20m^2 unit sizes are joined to build 1000m^2 platforms that are then interconnected to form 100x MW arrays. HDPE floaters are easy to connect and scale because of their light weight.

These structures have been tested in wind tunnels (the combined effects of aerodynamic loads and water motion have to be assessed). Vibrations can cause micro-cracks in the solar panels. The unit sizes of elastic membrane structures are in the range of 50-200m^2. Rigid platforms of rafts or pontoons for offshore deployment are available in much larger unit sizes of up to 10,000m^2. HDPE and elastic membranes offer lower cost than the larger metal platforms.

Fixed anchors are used for inland reservoirs, with concrete blocks for deadweight anchors chained to the floating platform. For shallower waters, steel piles are also used to pin the platform. Offshore platforms employ multiple cables with taut and flexible tensions to absorb the impact of waves, limiting platform drift with the tides.

Catenary mooring lines hang in a curve between the floating platform and the seabed, providing a restoring force to steady the structure for rising and falling water levels. The heavy, hanging weight of the lines acts like a spring that can pull the platform back in place after heavy waves.

Submersible cables are connected in a carefully planned slack to adapt to floater platform movement and the shoreline, to avoid torsional fatigue, mechanical wear and tear. Typical layouts are a lazy-wave, aka lazy-S, containing a buoyancy structure for the central portion, or for shallower waters a smooth catenary J-curve. Inverters are often grouped and placed on separate floaters for easy maintenance access. Built-in walkways are an important design consideration for O&M of the whole array.

Design knowledge from offshore wind farms is useful for this fast-growing type of solar installation. Further cost reductions with increasing economies of scale, materials improvements, e.g., polymer composites with a higher strength to weight ratio than HDPE at similar if not lower cost, and innovative designs, e.g. the integration of sunlight tracking, are in progress.

Environmental testing has shown that the floaters are not only compatible with but help aquaculture (though the plastic materials have to be carefully tested for leaching risk). On inland water reservoirs, the quality of water is improved by reducing the growth of undesirable algae.

Economical mounting structures for almost all types of surfaces exposed to the Sun, makes solar photovoltaic power a globally ubiquitous resource. Mounting design is instrumental to the modularity and scalability of PV systems.

--*--

Ch7 references, annotations, and supplementary reading

1. Influence of Photovoltaic Module Mounting Systems on the Thermo-mechanical stresses in Solar Cells by FEM Modelling
2nd European PV Solar Energy Conference and Exhibition, 20-24 June 2016, Munich, Germany
Andreas Beinert et al, Fraunhofer Institute for Solar Energy Systems
2. Fundamentals of Foundation Engineering and their Applications, G. Kouretzis
https://oercollective.caul.edu.au/fundamentals-of-foundation-engineering/
Understanding the theory and interpretation of SPT, CPT, pile load tests
3. Ground-screw manufacturer American Ground Screw:
https://www.americangroundscrew.com/applications/solar/
4. Pre-cast ballasts for ground-mounted projects:
https://www.terrasmart.com/products/ground-mount/ballasted/
5. Numerical study on the sensitivity of photovoltaic panels to wind load on array layout, Sustainable Horizons, 2024, Guangchen Jia et al.
https://doi.org/10.1016/j.horiz.2024.10010
6. Wind loading and its effects on photovoltaic modules: An experimental–Computational study to assess the stress on structures, Solar Energy, Andrew Wittver et al.
https://doi.org/10.1016/j.solener.2022.04.061
7. Wind Tunnel Experiments on Ground-Mounted Photovoltaic Solar Panels,
Revista Ingeniería de Construcción, 2019, A. Lopez et al.

https://scispace.com/pdf/wind-tunnel-experiments-on-ground-mounted-photovoltaic-solar-25lcjkemqw.pdf

8. Tracker manufacturers:
   Array Technologies Inc: https://arraytechinc.com/products/ - USA
   GameChangeSolar:
   https://www.gamechangeenergy.com/products-and-services- USA
   Nextracker: https://www.nextracker.com/- USA
   Arctech Solar: https://en.arctechsolar.com/solar-tracker/ - China
   Trina Solar: https://www.trinasolar.com/us/product/TrinaTracker - China
9. Rooftop mounting specialists:
   Esdec: https://www.esdec.com/en/ - Netherlands
   Arctech Solar and Schletter have introduced mounting structures using fibre reinforced polymer composites
   Schletter: www.schletter-group.com – Germany
   KB Racking: https://www.kbracking.com/projects - USA
   Chiko Solar: https://www.chikolar.com/products/ - China
10. Offshore solar photovoltaic potential in the seas around China, Applied Energy, Yi Wen and Pengzhi Lin
    https://doi.org/10.1016/j.apenergy.2024.124279
11. 1GW open-sea solar farm:
    https://sustainability-news.net/sustainability-news/china-launches-worlds-largest-open-sea-solar-farm-marking-new-era-for-offshore-solar/
12. An interdisciplinary literature review of floating solar power plants, Renewable and Sustainable Energy Reviews, 2025, Yujia Wei et al.
    https://doi.org/10.1016/j.rser.2024.115094
13. Solar floater suppliers:
    Sungrow Power: led with large-scale projects using HDPE floaters

https://en.sungrowpower.com/ProductsHome/21/floating-pv-system

AccuSolar: https://www.accusolar.com/our-design/, HDPE floaters – Florida, USA

Ciel & Terre: major manufacturer with an international portfolio of large projects - France

https://ciel-et-terre.net/omkareshwar-large-scale-high-challenge-fpv-project-india/

Ocean Sun: elastic membrane floaters – Norway

https://oceansun.no/projects/

Ch8. Rest of PV System

"Everything seems impossible, until it's done" – Nelson Mandela

PV system design, construction, and operations are interdisciplinary work from the factory production floor to project deployment and maintenance. To complete the electric power delivery system, we have to consider low voltage (LV) DC and low and medium voltage (MV) AC power cables, grounding and earthing, distribution boards, transformers and switchgear with integrated protective relays, sensors to monitor the system performance, and a SCADA for control, reporting, and coordination.

LV DC Power Cables

DC cables are the electrical lanes of a PV array, interconnecting the modules in series-connected strings to the inverters. Operating at LV, these cables carry high currents. The cables must be rated for a minimum of $1.25 \cdot I_{sc}$, i.e., 1.25 times the path short-circuit current, and maximum permissible voltage that is 1.2 times the nominal system voltage e.g. up to 1800V for a 1500V system. In the US, a higher cable current rating of $1.56 \cdot I_{sc}$ than the global standard is required. For two module strings running into a T-connector, $1.25 \cdot$ module I_{sc} for the two input cables and $2.5 \cdot$ module I_{sc} for the output cable.

The standard cable is copper (Cu) conductor in cross-linked polyethylene (XLPE) insulation. Aluminium (Al) is a cheaper alternative - but with much lower conductivity (35 mega-siemens/m Al vs 58 mega-siemens/m for Cu), and more vulnerable to damage from thermal expansion and contraction e.g., due to variable power supply, the environmental conditions. For lengths less than 100m: $4mm^2$ Cu XLPE size is recommended, 100-140m: $6mm^2$ Cu XLPE, and greater than 140m: $10mm^2$ Cu XLPE. Voltage drops can be calculated from $V = 2 \cdot I \cdot Length \cdot R$, where I is the current, length the distance the cable covers, R the resistance per unit length, and the factor of 2 for the 2-wire DC cables used in PV LV networks (current travels to the inverter load through the positive cable and returns in the negative cable). Take care that the DC cable polarities are correctly labelled and connected.

Apart from high conductivity, the cable must be capable of high thermal tolerance (short periods of 393K, continuous 363K), UV stability, anti-corrosion, and flexible mechanical strength. The Cu conductor is tinned or made in strands of bare metal (e.g., 56 strands for $6mm^2$), covered in single- or double-insulation XLPE dielectric, with a halogen-free, flame-retardant, and abrasion-resistant sheath (reduces toxic flames in case of a fire). Quality control checks of the cables are crucial to avoid:

- Inferior insulation: leading to short-circuits or ground faults
- Incorrect sizing: overheating and fire risk
- Cheap sheathing: cracks appear from UV exposure
- Moisture ingress: corrosion and insulation breakdown

During installation and maintenance, it is important to check for sharp bends, cable tray misfits, insulation damage from rodent bites, and discolouration from sub-standard materials.

High-purity electrolytic Cu (≥ 99.9% purity) is necessary for high conductivity. A thin tin coating to prevent oxidation is advisable for areas with hot and humid conditions. An XLPE dielectric strength of 120 kV/mm. LV cables don't normally have a semiconducting layer between the conductor and XLPE to smooth the electric field distribution (necessary for MV/HV DC cables). Electromagnetic energy transfer, given by the flux vector, flows between the conductor and the XLPE insulation at a velocity of ~2 • c/3, where "c" is the speed of light in vacuum.[1,2]

The electric field E around a cylindrical DC conductor of radius "a" and insulation outer radius "b", is given by $E(r) = V/(r \cdot \ln(b/a))$, where "r" is the outward radial distance, $a \leq r \leq b$, V the voltage between the conductor and sheath. E peaks at r = a. The thicker the insulation, the lower the peak electric field, and likelihood of dielectric breakdown leading to a dangerous energy discharge.

The magnetic field B around the conductor is given by

$B = (I \cdot \mu_o \cdot \mu_r)/(2\pi \cdot d)$.

"I": the current

μ_o: permeability of free space,

μ_r: the relative permeability of XLPE

"d": radial, perpendicular distance from the centre of the conductor

For an uninsulated conductor, as in overhead lines, the electromagnetic energy is not confined within the cable but flows around it in free space (an electromagnetic wave of power). Power loss in the cable is inversely proportional to its cross-sectional area: hence, larger cross-sections reduce the power loss but add to the cost.

Troubleshooting:

Electrical array faults in the DC cabling network are more problematic to trace than on the AC side, potentially leading to high fault currents damaging the inverters or arcs and fires. Poor DC cable quality, installation, or small animals chewing into the DC cables can cause it.

AC power flow faces significant cable inductance that self-limits current surges (L • dI/dt), but DC LV cables have a small inductance so fault current spikes just about instantaneously. AC faults tend to be localized, but DC faults rapidly propagate across the array. The PV module strings are wired in parallel – therefore a fault current in one string can flow backwards through the other strings, causing array-wide damage. A damaged DC cable may form a low impedance path to ground, drawing current not just from its own string of DC modules but from neighbouring modules as well. A DC ground fault current can have a low magnitude that escapes detection by overcurrent protection, but is sufficient to cause widespread problems. To sum up, DC fault currents are harder to isolate and control because of low inductance, no zero-crossover, and bidirectional flow in the LV network.

Time Domain Reflectometry (TDR) is a mature diagnostic tool for detecting DC cable faults e.g., broken cables, insulation breakdown, and loose connections.[3] There is no frequency signal to work with in DC power flow, so TDR is performed by injecting fast rising voltage pulses in the cable. If the cable has a fault, part of the pulse is reflected back. The time delay of the reflected pulse locates the fault. Larger reflection is indicative of a broken conductor, smaller reflections of moisture ingress or degrading insulation, intermediate reflections a sign of corroded or loose connections because of poor crimping. TDR is typically conducted offline, after carefully de-energizing the system, and isolating the cable to be tested. Spread-spectrum TDR can be carried out in-situ on operational, energized cables. TDR detects hidden faults, the proximate locations of the faults, but it is not a preventive or early detection technique. Magnetic field sensing of partial discharge in LV DC cables for early warning and preventive measures is worth further R&D, especially with the commercial development of vastly sprawling DC arrays in PV systems.

TDR is suitable for line-line and line-ground faults. For a ground fault, a low-frequency AC or audio signal can be injected. Tracing the suspected cable path with a handheld receiver wand (an inductive, capacitive, or audio sensor depending on the setup), can identify the ground fault location. Alternatively, in a more sophisticated approach, Hall Effect current sensors with differential Hall Effect detection, i.e. measuring the magnetic fields of the positive and negative conductors simultaneously, can pinpoint both fault and leakage currents. Magnetic field sensing can also be extended to partial discharge detection - widely used for higher voltage cables.

LV, MV AC cables

LV AC cables run from the inverters to the LV distribution board. Current (rms) rating follows the same 1.25 standard, i.e. 1.25 times the maximum output current of the inverter. The same thermal, UV, humidity tolerances apply as for DC, with a short-circuit current withstand of 523K for five seconds. For a 1500V system, the voltage rating is 1.8/3kV - 1.8kV_{rms} between a single phase conductor and ground, and 3kV_{rms} between any two phase conductors.

AC cables consist of a single-conductor Cu core or three conductor cores, in XLPE insulation. Three cores are preferred where there is shortage of space or for thermal management of unbalanced loads in 3-phase systems, while single-conductor cores provide redundancy and are easier for maintenance e.g., avoid the hardship of detecting line to line faults in three cores.

The conductors are in concentric strands to reduce the skin effect, with a semiconducting screen for the MV AC cable. The semiconducting screen smooths out internal electric field distribution to prevent insulation damage. For the MV AC cable from the distribution board to switchgear, current and voltage ratings are set by the short-circuit fault current and voltage safety margin assessment of the grid. A detailed ampacity calculation for the MV cable is prudent, due to the higher power load, to derate its temperature dependent current capacity for the actual operating conditions.

The DC and AC cables are a substantial part of the total system budget, at par with the inverters and mounting.

Earthing and Grounding Network

An earthing rod for the ground network is a bare Cu conductor, e.g. 16mm diameter, 1.5m depth, sized according to the formula for its resistance R_{rod}:

R_{rod} = (Soil resistivity/2π · rod length) (ln (4 · rod length/rod diameter) - 1)

R_{rod} should be less than 5Ω.[4] In case the soil resistivity is too high, chemicals can be added to the soil or sufficiently spaced earthing rods can be placed in parallel to reduce R_{rod} to the threshold value. The grounding network consists of copper conductors, e.g., 50mm^2 copper cables, connecting all the mounting frames and non-conducting metal parts to the earthing rod.

LV/MV Distribution Board

The LV/MV distribution board (DB) collects and distributes the AC power from the inverters or AC combiner boxes (for multiple inverters) to the transformer-switchgear, loads, isolating the different system sections for protection. **Miniature circuit-breakers (MCB)** protect the incoming AC cables to the distribution board. **Moulded circuit-breakers (MCCB)** protect the higher-current AC cable from the DB to the transformer-switchgear. **Surge protection devices (SPD)** are included: type 1 for upstream electrical transients and type 2 for induced current, voltage surges, or lightning residual impulse in the downstream circuit.

Assessing the distribution board features is an important step in retrofitting PV systems. Older distribution boards may require additional protective elements. The costs of the distribution board span a wide range, varying by the project size, detailed specs ~ $10k-$100k.

Step-up Transformer (LV/MV, MV/HV)

The LV/MV transformer is almost always the responsibility of the solar project, and occasionally so is the MV/HV transformer. Transformer specifications are broadly determined by a grid impact study typically carried out by the grid operator, compatibility with the grid network, and the planning of the grid connection operations (part of a grid connection agreement).

For the LV/MV transformer, **epoxy resin dry-type transformers**[5,6], or perhaps soon the emerging **solid-state transformers**[7] (10-100kHz high-frequency switching with AC/DC/AC power transistors), are preferable to the older oil immersed transformers for quite a few reasons:

- Safety of fire self-extinguishing resins
- Lower maintenance cost (do not have to monitor the oil, worry about oil leaks), sealed enclosures protection against dust and moisture ingress
- Environmental concerns: less adverse impact for end of project disposal
- Select class-H or class-F insulation for dry-type for elevated temperatures, options of better ambient temperature operating ranges than oil-immersed counterparts
- Solid-state transformers (SST) are a new option: compact designs, with distinct capabilities of harmonics mitigation and voltage regulation

Select transformers with a higher K-rating, attention to the mitigation of harmonic currents:

- Higher K-ratings for thermal tolerance of harmonic currents
- Winding configurations to cancel harmonic frequencies, and reduce eddy currents
- SSTs actively manage and eliminate harmonics

Additional considerations in selecting transformer:

- Magnetic core selection e.g., ultra-thin ribbons (nanocrystalline) of soft magnetic FeSiB-NbCu alloys offer high efficiency and low losses
- Optimised design for both no-load and load losses
- Advanced monitoring capabilities, connected to SCADA

SST might take some time for full commercial maturity at competitive prices, but dry epoxy resin type is a safe, reliable option with many vendors to choose from.

FAT oversight of the transformers is critical, to ensure it meets the agreed datasheet specifications - transformers are made to order, not an off-the-shelf product. Transformer warranties are typically only for one to two years.

For MV/HV, oil immersed transformers are dominant. Synthetic esters for the oil are a favourable choice, supplemented by forced cooling e.g. OFAF (oil forced, air forced over radiators), OLTC (online tap changers) for voltage regulation, dissolved gas analysis, and advanced real-time monitoring especially for partial discharge.

Switchgear

The switchgear works in tandem with the step-up transformers as the interface between the PV system and the grid. The specifications are again largely determined by the grid impact study requirements. Preferably, buy the transformer(s) and switchgear from the same manufacturer as an integrated package. Switchgear consists of heavy-duty circuit-breakers, power distribution busbars, isolation disconnectors, protective relays, and auxiliary items like an uninterruptible power supply (UPS). The protective relay protects against overcurrent, voltage and frequency deviations, a list of faults specified in the grid code or specific grid connection agreement for the project. Programmable microprocessor protection relays are now the norm. Partial discharge sensors,[8] for continuous real-time monitoring, enable early warning against dielectric breakdowns. Consider integrating switchgear monitoring with the SCADA.

Ring Main Units (RMU) provide a convenient type of compact, modular, metal-enclosed, and self-contained switchgear for MV connections. The power flows through a circular loop of switches and protection – there is no alternative path for the power flow, except re-routing through the ring. Faults can be quickly isolated and separately managed, without necessarily cutting off all downstream power flow.

Sulfur Hexafluoride (SF6) has been the main choice for the electrical insulation medium in the switchgear isolation disconnector, but other options have become available. SF6 can decompose in electrical discharge, arcing, releasing toxic, hazardous substances. Clean air, aka synthetic air,[9] has comparable electrical arc quenching effectiveness without any toxic byproducts. Vacuum interruption technology[10] is another option.

FAT supervision of switchgear, always a customized product for the grid connection, is essential e.g., for confirming switchgear drawings match the final product, circuit-breaker tests (open/close timing), insulation system validation, manual override function, interlocking mechanism, tests for leaks and partial discharge, among other specifications. Step-up transformers and switchgear are not just required for PV, but all utility-scale electric power sources.

Performance Sensors

PV modules are rated at STC. The actual operating conditions seldom, if ever, match the STC. In addition, we observed that module performance is a projection of laboratory accelerated tests with assumptions relating the accelerated periods to actual operating time. Therefore, it's necessary to calibrate the actual performance data to the expected. For this purpose, a minimum of two types of sensors are necessary: irradiance sensors that measure the actual irradiance in W/m^2 and irradiation (time integral of the irradiance for different periods: kWh for that period), and temperature sensors for the module temperature.[11] In addition, a weather station including an ambient temperature sensor, irradiance sensor (for global horizontal irradiation: GHI), and anemometer (for wind speed) is required – certain assumptions will have been made in how the PV array temperature is affected by ambient conditions, so these assumptions have to be validated.

For the irradiance sensors on the module array, the advisable choice is a silicon photodiode rather than a pyranometer, assuming the PV system comprises silicon solar modules. The reason is to avoid spectral mismatch in calibrating system performance. Minimum specs for the silicon diode:

- Spectral range of 350-1100nm
- Measurement range of 0-1500W/m^2
- Linearity error less than +/- 1% over the measurement range
- Tilt response of +/- 1% at 60 degrees
- Spectral error of +/- 3%

There should be at least one irradiance sensor for every sub-array orientation. For large arrays, the irradiance sensors should be appropriately placed to obtain a reasonably accurate measurement of the irradiance over the whole array. The irradiance sensors should be mounted in the plane of the array for Global Tilt Irradiance (GTI) measurement.

Platinum Pt1000 temperature sensors (platinum resistance changes with temperature) are to be mounted at the back of the module, away from the junction boxes. The Pt1000 sensors are to be distributed over the array to give a reasonably accurate estimate of the module array temperature. The minimum spec for the Pt1000 Class A sensor:

- 233K to 373K temperature range
- +/- 0.3K to +/-0.55K accuracy over the temperature range

- Response time of less than 3s in 1m/s air flow
- Less than 0.05K drift per year
- Self-heating error less than 0.1K at 1mA

A few co-located silicon diodes and Pt1000 sensors are advisable, i.e., on the same module.

These measurements are sent to the SCADA for calculating the aggregated "Performance Ratio (PR)" of the entire module array according to the selected PR definition – e.g., a frequently used PR_{STC} definition adjusts for the differences between STC and actual operating conditions.

PR_{STC} adjusted for irradiance and temperature differences from the STC is given by the ratio of two terms:

The numerator term is $\sum_k P_{out,k} \cdot t_k$.

P_{out}: power output measured in the period k

t_k: duration of the measurement period k

$(P_{out,k} \cdot t_k)$ is summed over "k" periods

The denominator is $(\sum_k C_k \cdot P_0 \cdot GTI_k \cdot t_k)/1000W/m^2$.

P_0: the STC power output of the installed modules DC W_p capacity for the same measurement time period t_k as in the numerator

GTI_k: global tilt irradiance on the array measured in period k

$1000W/m^2$: the STC irradiance

$C_k = 1 + \gamma \cdot (T_{module,k} - 298K)$.

γ: the relative maximum temperature coefficient of power loss by temperature for the module

T_{module}: the module array temperature in Kelvins (K), since we are using 298K, measured in time period k

t_k: the same duration of the measurement period k as in the numerator

Denominator is summed over the same k periods as the numerator

The information of Ɣ is provided by the module manufacturer: validate the Ɣ value in FAT and then check it in field-tests, e.g., -0.0025/K for an HJT module (negative 0.25%/K).[12]

The "k" periods should add up to at least a summation block of fifteen minutes for a unit PR_{STC} calculation.

This PR_{STC} can be calculated in the SCADA from the sensor data that is transmitted to it. The rules for calculating PR_{STC} are programmed in the SCADA.

SCADA

The SCADA consists of hardware (industrial grade computer, communication devices) and software to control the inverters, and any other integrated devices e.g., the transformer and or switchgear, by a Power Plant Controller (PPC) unit, collect all the sensor data, use it to calculate irradiance and temperature adjusted performance data, display the monitored system information to permitted users.[13]

SCADA functions include:

- Active power P control from the inverters
- Ramp the inverters up and down, shutting down if required

- Q(V) control from the inverters
- Adjust performance setpoints for strings
- Calculate PR metrics for whole system, strings
- Display system and inverter level performance
- Line voltages, currents for each string
- Display irradiance, temperature data
- Detect abnormalities, convey alarms to users/plant operators

The minimum technical specifications for the SCADA to perform these functions:

- Support Modbus TCP/IP, RTU protocols
- Ramp rates for inverter control meet or exceed inverter capability < 100ms
- Data logging time interval < 15s
- Alarms response < 1s
- Sufficient ports for inverters and performance sensors, including weather station ambient data
- Cloud data backup, cybersecurity for stored and shared information
- User interface to receive commands from users – preferably Web-based interface for remote access
- Programmable key performance indicators, reporting

SCADA enables project managers to program sophisticated rules for the analysis of the PV system, set up customized alarms and periodical reporting. It is used to supervise and control the inverters - how effectively it is utilised depends on the diligence and ingenuity of the project managers. Data collection and analysis logs provide a technical due diligence track record of the solar asset. A PV system that does not show intelligent utilisation of the SCADA is unlikely to have been well maintained.

We have covered the fundamentals of all the main components of a PV system. There are other peripheral items to budget for e.g. the fencing of the project site, site lighting, security surveillance cameras, an auxiliary transformer, a warehouse for storing the received goods, but these items are not of major importance requiring an in-depth understanding to plan a solar project.

--*--

Ch8 references, annotations, and supplementary reading

1. Energy flow from a battery to other circuit elements: Role of surface charges, Manoj K. Harbola
DOI: 10.1119/1.3456567
For a detailed quantitative discussion of DC energy flow along a cable
2. Speed of electricity transport by a power cable:
https://en.wikipedia.org/wiki/Speed_of_electricity
Clarifies misconceptions about how electricity travels. The cable acts as a waveguide for the power flow. Shows how 2 • c/3 is applicable.
3. An Overview of Spread Spectrum Time Domain Reflectometry Responses to Photovoltaic Faults, Mashad Uddin Saleh et al.
https://www.osti.gov/servlets/purl/1659850
4. 142-2007 - IEEE Recommended Practice for Grounding of Industrial and Commercial Power Systems
DOI: 10.1109/IEEESTD.2007.4396963
5. Basic overview of dry-type transformers from an industrial supplier:
https://www.chintglobal.com/global/en/about-us/news-center/blog/cast-resin-dry-type-transformer.html
6. Resource library on epoxy resin dry-type transformers from another:
https://jstpower.com/product-literature/cast-resin-dry-type-transformers/
7. Solid State Transformers: A Comprehensive Review of Technology, Topologies, Applications, Research Gaps, and Future Directions, 2025,
Journal of Power and Energy Engineering, D. Chandari et al.
DOI: 10.4236/jpee.2025.136003
8. Partial Discharge sensor from Megger:

https://www.megger.com/en/promotions/sa-PDSCAN

9. Synthetic Air Switchgear from Lucy Electric: https://www.lucyelectric.com/all-secondary-switchgear-products/ring-main-units/
10. Vacuum interruption technology switchgear from Eaton: https://www.eaton.com/us/en-us/products/electrical-circuit-protection/vacuum-interrupters/vacuum-interrupters---fundamentals-of-vacuum-interrupter-technol.html
11. IMT Solar: www.imtsolar.com – Germany

    Specialises in irradiance and temperature sensors for PV systems
12. The STC temperature corrected (adjusted) PR is defined in IEC 61724-1. All PR formulas adjust for irradiance differences from the STC. Temperature adjustment should be applied for time periods of less than a year to cover all the seasonal variations of operating conditions. The PR formulation to measure system performance evaluation should be the same as what was used for simulating the projected system performance – e.g. if PVSYST software simulated the designed system performance, then its PR definition is applicable: https://www.pvsyst.com/help-pvsyst7/performance_ratio.htm
13. SCADA vendor examples:

    Meteocontrol: www.meteocontrol.com – Germany

    Green Power Monitor: www.greenpowermonitor.com – owned by DNV, Norway

Ch9. Project Planning and System Design

"Plans are worthless, but planning is everything" – Dwight D. Eisenhower

Any place with a surface exposed to the Sun, where a solar panel can be mounted, has a measurable potential to produce electricity through the PV process. The electricity can be consumed at the point of generation, shared, or sold to others. Commercial electric power producers seek to generate solar electricity in bulk and sell it to others. Project planning considerations are considered from the dual perspectives of a project developer (who intends to initially own the solar asset) and consumer, who respectively sell and buy electricity through an AC grid network. Electricity has to be injected into the grid according to its code for onward transmission to the grid electricity consumers. The electricity purchasing agreement can be between the project developer/asset owner and a grid operator, the government, a private corporation, or a community of consumers.

Real Estate: Land and Grid

There are two types of real estate the project developer has to secure: the land for constructing the project, and space on the electric grid i.e., a grid connection. We include rooftops and accessible water surfaces in our general references to land. A substantial ground-mounted project is the reference case for preliminary discussion. Just about all unobstructed land is useful for PV production, but some of it is more suitable than other locations. From the project developer's point of view, they want land that receives greater annual irradiance but costs less to install the system. In addition, proximity to the grid is highly desirable to save on longer power cabling costs. The grid operator, on the other hand, has their own priorities: how the power plant will fit in the scheme of balancing active P and reactive Q power on the grid, selling the electricity to their customers for minimum cost and maximum profit. With these considerations in mind, the grid operator allocates or permits grid connection points. Since electricity can be quickly transported over long distances, at close to the speed of light and dynamically controlled across a sprawling network, the grid operator has a considerable degree of flexibility in deciding where to permit grid connections.

The project developer has to devise a technical plan for maximising solar energy yield from the available land, and satisfying the requirements of the grid operator to secure a grid connection. The grid operator will want to know the electricity generation profile the project developer intends to sell, to grant a grid connection point with a certain power evacuation capacity subject to their grid code. Technical planning has to therefore start with knowing what the land can produce, but also a qualified approximation of what the grid operator might accept. There is a lot more land and landowners available in any region, than grid lines and grid operators, so the grid gatekeeper is the main determinant of what project size to pursue.

Technical Planning of Project Proposal

To assess the technical suitability of land, aside from legal diligence and securing it for the project duration without any encumbrances, the project developer has to undertake the following assessment:

- Topography: shape and contours of the land
- Soil composition, mechanical properties: load bearing strengths, through standard penetration test (SPT), essentially seeing how many blows are needed by a standard instrument to penetrate, cone penetration test (CPT)
- Boreholes ~ 10m deep, depending on what one discovers
- Chemical analysis of soil for any corrosive properties, hazards
- Electrical resistivity measurement of soil – important for earthing (re: previous chapter formula for earthing rod)
- Geological report about the site area: insights into the type of soil
- Seismic data
- Flooding and water logging

There are two objectives of this information:

- Identification of any red flags, that make the land unsuitable
- Information for the mounting designer/manufacturer

In addition to this information, the project developer has to let the mounting designer know of the modules they plan to use, static and dynamic wind tolerances. The mounting manufacturer makes a preliminary assessment of what type of mounting structures are suitable, aiming for a minimal cost of civil works to prepare the land for the selected design. Civil works generally involve removing obstructions, levelling, and grading the land to the extent necessary for the proposed mounting structure. A PV array layout is prepared on the basis of it. We have seen (Ch7) that trackers deliver much higher energy yield, and hence revenue +15-25% per year, for a small upfront incremental cost < 3% of the capex budget. The first step is to therefore see whether trackers are feasible. If major civil works are being planned, it should be for the purpose of enabling trackers. If trackers are not feasible, the next step is to see how the fixed-tilt array layout can be optimised for maximum energy generation - a matter of not only maximising the number of modules to fit in, but also in what orientation for the incident site insolation.

Push-Pull Test

Once the general array layout has been agreed, the mounting designer will want push-pull tests conducted for the mounting foundation design they have in mind e.g., pile foundations or ground-screws.[1] It is a good idea to let the mounting designer conduct their own push-pull tests, or convince them of it, if they are located nearby. In any case, have the push-pull tests done to the mounting designer's satisfaction.

Push-pull tests involve ramming a steel pile or ground-screw into the ground, applying downward, lateral, and uplift forces in steps to the specified design limits:

- Axial load capacity in compression test; hydraulic jack with gauge, load cell for direct force measurement
- Displacement in compression test: linear differential sensor
- Torque: gauge displacement during rotational force of insertion, rotational force measurement to check when load twists
- Pull test for uplift resistance, heave displacement
- Lateral force: bending displacement, inclinometer for deflection angle
- Cyclic loads to mimic dynamic wind stress

A load displacement curve is obtained for the test sample showing the elastic and plastic ranges (recoverable vs permanent deformation), and ultimate load limit. The test results reveal information about:

- Ultimate capacity: max load before destruction
- Allowable capacity: safety margin applied to ultimate capacity for each force limit i.e., lateral, rotational bending, axial compression
- Pile-soil friction: tension resistance for pullout

A project without mounting design, based upon comprehensive push-pull tests, is at risk. These tests are not expensive, providing greater surety about the lifetime mechanical stability of the project. Unstable mounting may vibrate, shake, under severe dynamic loads, or lose alignment over time, from the static loads only – leading to degradation mechanisms, underperformance from the projected energy generation profile. Beware of solar projects deploying mounting and array designs, that are not based upon such tests!

In parallel to the push-pull testing, the project developer shall want to estimate the expected energy yield for their array layout. Energy generation simulation has to be carried out. Standard commercial and free open-source energy simulation software is available for the purpose. The module and inverter manufacturer will provide their product data in a file format compatible with the software package (it is the technical data shown on the module/inverter datasheet, so it can also be manually inserted with reference to the datasheet).

Strings Configuration

The LV design involves planning the series connected strings of modules for the MPPT inputs of the inverters. Inverters can typically support two strings per MPPT for specified limits of the input current and voltage. Strings connected in series, adding up the module voltages, and parallel to a MPPT, summing up the currents of different strings, is called the series-parallel configuration. PV energy simulation software calculates the minimum and maximum string lengths, but worthwhile to also check it manually. Solar cell junction voltage varies linearly with temperature (Ch5) - hence, the string-level voltage of modules rises in colder temperature and vice versa for hotter.

For the maximum string length, the open-circuit voltage of the string should not exceed the inverter input voltage limit at the coldest expected temperature:

$N_{max} \cdot V_{oc} (T_{coldest}) \leq V_{max,\ inverter}$.

N_{max}: maximum number of modules for the string length

V_{oc} ($T_{coldest}$): module open-circuit voltage at the coldest expected temperature

$V_{max,\ inverter}$: the maximum input voltage of the inverter

If the string is too short, its maximum power point voltage at the hottest expected temperature will be too low to turn the inverter on. So, for the minimum string length:

$N_{min} \cdot V_{mpp} (T_{hottest}) \geq V_{min,\ inverter}$.

N_{min}: minimum number of modules for the string length

V_{mpp} ($T_{hottest}$): module maximum power point voltage at the hottest expected temperature

$V_{min,\ inverter}$: minimum input voltage to turn the inverter on

The calculation process is:

Estimate the PV cell junction temperature T_{cell} range from the coldest and hottest expected ambient temperature $T_{ambient}$:

$T_{cell} = T_{ambient} + (NOCT - 293K) \cdot 1{,}000W/m^2/800W/m^2$.

$T_{ambient}$: the ambient temperature

NOCT: the normal operating cell temperature given on the module datasheet for $800W/m^2$ irradiance, e.g., a HJT value of 315K.

Insert the coldest and hottest expected ambient temperature from the meteorological reference source for the site to obtain the coldest and hottest expected cell junction temperatures: $T_{coldest}$ and $T_{hottest}$ respectively.

Note the cell junction temperature does not only depend upon the ambient temperature but also the incident irradiance.

$V_{oc}\ (T_{coldest}) = V_{oc,\ STC} \cdot [\ 1 + (ß\ of\ V_{oc}) \cdot (T_{cell,\ coldest} - 298K)]$

$V_{mpp}\ (T_{hottest}) = V_{mpp,\ STC} \cdot [1 + (ß\ of\ V_{mpp}) \cdot (T_{cell,\ hottest} - 298K)]$

Where ß of V_{oc} and ß of V_{mpp} are provided by the manufacturer: the coefficients of V_{oc} and V_{mpp} variation with temperature from STC, e.g., ß of V_{oc} = -0.0021/K, ß of V_{mpp} = -0.0027/K.

$N_{min} = ceil\ (V_{min,\ inverter}/V_{mpp}\ (T_{hottest}))$

$N_{max} = floor\ (V_{max,\ inverter}/V_{oc}\ (T_{coldest}))$

$T_{coldest}$ and $T_{hottest}$ are the calculated cell junction temperatures for the expected ambient limits.

Practically, the system designer shall choose the N_{max} string length to reduce the system cost (most efficient use of inverters and minimises cabling cost).

The string current is also checked for the inverter input current limit:

$I_{string} = I_{sc,\ STC} \cdot [1 + (ß\ of\ I_{sc}) \cdot (T_{cell,\ hottest} - 298K)]$

I_{string} the string current increases with temperature from Isc, STC, the current at STC, according to the ß of I_{sc} coefficient of current variation from STC shown on the module manufacturer datasheet.

Note: a customer who does not have access to energy simulation software can use these equations to check their installer has prudently, and not extravagantly, selected the string lengths for the installation. (For a rooftop in a hot climate, $T_{ambient, hottest}$ should be estimated for a hot rooftop rather than simply relying on meteorological data for the ground)

If the project developer is running short of MPPTs, they can connect strings in parallel to a T-connector[3] (MC4 branch connector) for one MPPT. Have to make sure the string currents do not add up to more than the inverter input current limit. The strings connected to the T-connector should be equal in length, or the MPPT will not be able to optimise the MPP for both strings (also holds for strings connected to the same MPPT without a T-connector). The T-connector should be a last resort, because it makes fault monitoring more difficult, and requires adjustments in the SCADA for performance calculations e.g., for the PR per string. Novel string configurations like total cross-tie connections, honeycomb total cross-tie, and a quadrant swapping technique for series-parallel strings are being explored. Superior system power performance under partial shading of modules is the motivation. Cloudy conditions, haze, or dirt, can all lead to partial shading or more generally non-uniform irradiance across the strings.

Energy Yield Extrapolation

The module, inverter performance specification data, strings configuration, and tilt-azimuthal orientations of each string, is entered into the energy simulation software. The software requires a shading model input to calculate shading losses (more important in a built environment), inputs for soiling, module mismatch, cabling resistance, transformer losses - loss inputs that the project developer has to estimate. There is also an assumption of grid availability: a contingency loss attributed to its unavailability. The software asks the project developer to choose a meteorological database and reference period to construct a typical meteorological year (TMY) for the simulation - TMY inputs of hourly irradiance and temperature values to simulate the whole year performance. The software then performs these calculation steps for the energy yield of the system in the first year of operation:

System energy yield at STC:

Energy yield of the module DC W_p capacity of the system in DC kWh at STC.

Module power correction for simulated operating conditions:

Adjustment of system energy yield in DC kWh, for irradiance, temperature, and angle of incidence variation from STC.

Near and far field shading effects on the module array.

Soiling losses, typically a judgement call for the project conditions.

Mismatch losses (minimizable in practice by careful sorting).

BOS energy losses:

Debits energy losses for inverter efficiency of power point tracking, DC to AC conversion, cable resistive losses, and transformer losses.

Final simulation result: energy yield in AC kWh for the first year of operation, (AEY1), annual kWh/kW$_{p\text{-}AC}$, PR calculations, with a monthly break-up.

The project developer then applies the module annual power attenuation x% (e.g., 0.25%/year for a HJT module), specified by the module manufacturer, to calculate the annual energy yields for years 2 to 30 (assuming the module has a thirty-year power performance profile):

AEY2 = AEY1 • (1 - x%), AEY30 = AEY1 • (1 – 29 • x%),

i.e. AEYn = AEY1 (1 – (n - 1) • x%)

Where AEYn is the annual energy yield in year n, with n= 1 to 30.

The project developer may choose to add a safety margin on top of the module manufacturer's annual power attenuation rate e.g., change 0.25%/year to 0.5%/year. The AEY, when normalized by the AC W_p capacity of the power export connection, is reported in kWh/kW$_{p\text{-}AC}$, and sometimes called the specific production capacity of the PV system. Grid operators specify the available AC W_p capacity for the PV project planners.

Grid Connection Cooperation

This first iteration of a system design, often called the conceptual design, forms the technical proposal to the grid operator. The grid operator uses this system design information, including necessary grid code requirements of the transformer and switchgear specifications, to conduct a grid impact study.

The grid impact study can only be conducted by the grid operator or their selected agency for it, because only they have the necessary information about their grid network and operating protocols. The grid impact study is typically shared with the project developer for discussion and any necessary adjustments. The project developer is usually charged by the grid operator for the grid impact study. In some cases, the grid operator may ask the project developer to have the grid impact study conducted themselves, but in such a situation the project developer has to be provided with all the requisite information by the grid operator. Once the grid impact study is approved, the project developer can start working on a final design and grid connection agreement with the grid operator.

The grid connection agreement (GCA) is a critical technical and legal document for the project.[2] Technical oversight of each step of the GCA negotiation, drafting, and finalization is essential. It should not be compartmentalized and passed on to the lawyers for their purview and work only, after the conceptual system design and grid impact study have been approved. The implementation of the GCA requires close collaboration from both sides. PV systems are coupled to their grid networks, unable to control faults arising on the other side. The system switchgear is meant to protect it from external faults but there are limitations. Faults arising from the grid network may disrupt the PV system power flow, compromise power quality (e.g., harmonics from non-linear loads on the grid-side), and damage equipment. Changes in the grid network, beyond the control of the solar asset management, might adversely affect the PV system operation. Joint responsibilities for maintaining the agreed operational parameters of the grid point of connection (POC), and point of common coupling (PCC), have to be established in the GCA for the project duration.

The GCA, or by default the grid code, specifies the active and reactive power controls, responses to frequency and voltage deviations, fault ride-throughs, low and high earth fault limits, overcurrent protection, anti-islanding, SCADA communication and reporting, the PV system must satisfy. It might include P and Q control as a function of PF, or an independent Q(V) priority. The project developer ensures these technical conditions are incorporated in the final design, in particular with its SCADA and switchgear suppliers.

Financial Planning and Commercial Negotiation

There is a parallel commercial negotiation process between the project developer and the grid operator, or other electricity buyer (for the project output offtake), for a Power Purchasing

Agreement (PPA). There are numerous variations possible for a PPA, including the following and hybrid combinations thereof:

- Fixed tariff in $/kWh for all electrical energy delivered at POC
- Fixed quantity of electrical energy (kWh) at fixed tariff
- Minimum guaranteed kWh at fixed tariff, with a penalty for shortfall
- Fixed quantity of kWh at inflation-index linked tariff
- Minimum guaranteed kWh at inflation-index linked tariff
- Tariff indexed to gas price for fixed quantity or minimum guaranteed kWh
- Contracts for Difference (CfD) tariff: hedged to market clearance price
- Time of Use tariff: payment varies according to time tiers
- Merchant PPA: market price of $/kWh for all electricity needed at the time e.g., market price set by wholesale market clearing price
- Virtual PPA, essentially a corporate CfD: the corporate buyer agrees a strike price hedge with seller, and pays the difference if market price < strike price, receives rebate if market price > strike price
- Sleeved PPA: seller is paid for electricity “sleeved” by the grid operator to a remote buyer, the buyer also pays the grid operator for the sleeving

In all cases, the project developer has to calculate a project lifecycle selling price - the levelized cost of electricity (LCOE) for the expected system lifetime. For context, we shall discuss the LCOE calculation, and related commercial negotiations, for the frequent business-case of minimum guaranteed energy delivery at a fixed tariff to the grid operator (electric grid utility), with a penalty for shortfalls. Why would the utility demand a minimum guaranteed energy delivery? It is reasonable to assume there is more than one project developer in the market who wants to sell electricity to the utility. The utility is allocating the grid connection to the project developer on the expectation that they will deliver a minimum guaranteed energy schedule for the PPA duration. If they don't, the utility incurs an opportunity cost and has to pay other sources for the shortfall (most likely higher spot prices than the fixed solar PPA tariff).

LCOE

We assume the project developer, seller A, is in competition with another seller B for the grid connection and minimum guaranteed energy schedule, fixed tariff contract with the grid utility. The utility has fixed the available grid connection capacity in $MW_{p\text{-}AC}$ - (the AC subscript denotes AC watt-peak power export capacity), say $10MW_{p,\text{-}AC}$, for the technical design and grid impact study validation. We shall look at the financial planning and commercial negotiations, from both the viewpoint of the solar electricity seller and the grid utility buyer.

It is important to note that the LCOE in practice is not a cost but actually the electricity price in $/MWh (the tariff) offered to the buyer.

The standard LCOE definition is:

LCOE = total lifecycle energy costs/total lifecycle energy generation

The numerator includes not only the seller's cost of project works, hardware and software, but also their financing cost and profit expectation. The profit expectation of the seller is the internal rate of return (IRR) on their equity.

The denominator of the total lifecycle energy generation, the simulated energy generation profile is given by

Total $E_n = \sum_{n=1}^{n=30}$ AEY1 MWh • [1 - annual attenuation x% • (n - 1)]

Where "n" is the year, AEY1 MWh is the AC MWh delivered to the grid point of connection in year 1 (calculated in a whole system energy production simulation), x% is the annual attenuation rate, E_n is the energy delivery in year n, and total E_n is the aggregated energy delivery over n years on the left-hand side. AEY stands for Annual Energy Yield in kWh/year or MWh/year.

We first consider the situation where the seller is financing the total upfront project cost, TC, out of their own pocket i.e., 100% equity financed. For the project to be worthwhile to the seller for a fixed tariff of LCOE, and the time value of their money, $\sum_{n=1}^{n=30}$ (LCOE • E_n - OM)/(1 + IRR)^n=TC, where OM is the total running cost each year including technical operations and maintenance, annual land lease payment, and annual administrative costs including recurring insurance policies.

To solve for LCOE:

$\sum_{n=1}^{n=30}$ (LCOE • E_n / (1 + IRR)^n - $\sum_{n=1}^{n=30}$ OM/(1 + IRR)^n = TC

LCOE $\sum_{n=1}^{n=30}$ E_n / (1 + IRR)^n = TC +$\sum_{n=1}^{n=30}$ OM/(1 + IRR)^n

LCOE = [TC + $\sum_{n=1}^{n=30}$ RC/(1 + IRR)^n]/ $\sum_{n=1}^{n=30}$ E_n / (1 + IRR)^n.

We make some reasonable estimates based on contemporary prices for TC, OM, and $E_{n,}$ for the illustration of a 10MW$_{p\text{-}AC}$ project - the grid operator has assessed in a grid impact study its grid connection fits 10MW of AC power for STC rated power delivery of the PV modules and the simulated system performance:

Project Summary		10MWp-ac 30-year PPA	
Bill of Quantities (BOQ)	% of Total	DC STC Wp	10MWp
Module	31%	AC capacity	10kVA
Mounting	13%	DC STC Efficiency	24.5%
Inverter	6%	GHI	2MWh/y.m2
Tr-swg	13%	Tilt Gain	8%
Cables	9%	GTI	2.16M"h/y.m2
SCADA(PPC)	3%	STC DC to AC grid POC loss	12%
Development	6%	Collector Area	40816m2
Installation	14%	AEY-1	19008MWh
Spares	4%	Anuual attenuation Y2-Y30	0.30%
Contingency	3%	Total O&M cost p.a. fixed	$150,000
Total	$3.2 million	Land Lease 8ha, admin, insurance	

The land terrain is too difficult for trackers - civil works to grade the slopes sufficiently would cost too much. Natural slopes suit a fixed-tilt project with a 8% tilt energy gain in the plane-of-array irradiance, aka Global Tilt Irradiance (GTI), from the Global Horizontal Irradiance (GHI).

GHI is the site irradiance measured on a horizontal plane - the data is imported from meteorological databases.

$GHI = DNI \cdot \cos \theta_z + DHI$.

DNI: direct normal irradiance component to the horizontal plane, in a direct beam path from the daily movement of the Sun

θ_z: zenith angle of the Sun to the horizontal plane

DHI: diffuse horizontal irradiance

The project is in a top-tier insolation area of GHI = 2MWh/year per m^2. GTI for the array orientation gain = 2.16MW/year. Since the array is closely mounted on a sloped terrain (little room for reflected light from the land), bifacial modules are not used and there is no albedo gain.

Seller A assesses and guarantees a 1% annual power attenuation drop in year 1 and linear 0.3% per annum attenuation profile from year 2 to year 30. The total energy system losses for AEY1 are limited to a best-in-class 12%, including the first year module power attenuation drop of less than 1% (module power attenuation changes from year 1 to year 2).

The system takes exactly one year (year-zero) to build and commission (a reasonable time period for this project size on a difficult terrain), and for simplicity, we assume the entire cost is paid at the start of the year - this is sometimes called the overnight cost for such financial modelling. The overnight cost assumption works better for energy projects with short construction periods - as is the case for solar, but certainly not nuclear or coal power plants. Annual revenue is paid in a lump sum at the end of each year - in practice, it is normally a monthly or quarterly payment. The entire expenditure is rolled up in one figure for the project works with a construction gap of one year between the total cost expenditure and first annual revenue payment - i.e., a lapse of two years from total cost expenditure to initial revenue generation. These simplifications do not affect our analysis, as long as we consistently apply it to the ensuing comparisons. When project developers contract the project works to an EPC (engineering, procurement, and construction company), typically ~5% of the total cost is spent over two years after the project has been built and started generating revenue – i.e., in the first two operational years post the Provisional Acceptance Test at the contractually agreed Commercial Operation Date (the project developer covers the technical construction risks through a combination of the payment schedule and performance security).

We assume the project developer is both the complete investor - the project is 100% equity financed - and EPC. The IRR for an unlevered project, i.e., with no third-party debt financing, is called the Project IRR. The seller estimates the Project IRR for each LCOE price they can offer to the buyer – it seems obvious a higher LCOE (charged price per MWh) is desirable for the project developer to reach a higher IRR. An upper bound on the LCOE comes from the alternatives available to the buyer i.e., prices offered by other power sources. The lower bound is set by capex cost and profit expectation, including a cutoff year for when revenue falls below fixed annual O&M cost (O&M could be adjusted to carry on, to the extent possible). Project IRR is benchmarked against the returns of simply keeping the same amount of money in a fixed-income deposit.

For our $10MW_p$ example, seller A finds:

LCOE $/MWh	Project IRR	10MWp project illustration
10	-6.6%	100% equity financed
20	5.0%	Summary data as above
30	11.0%	
40	15.9%	
50	20.3%	
60	24.5%	
70	28.4%	
80	32.1%	

Note the differential rate of change of Project IRR with LCOE - d(Project IRR)/d(LCOE) - varies non-linearly. It reaches a maximum between $20/MWh and $30/MWh at about 0.83% per $/MWh.

The LCOE price the seller offers to the buyer depends on their profit and associated Project IRR expectations - a seller with a lower IRR expectation is satisfied by a lower LCOE. If a competitor to seller A came along with exactly the same system performance and project costs but a lower Project IRR expectation for satisfactory profit, they would take the project away.

The LCOE dependence on financial considerations becomes clearer once we factor in the cost of financing capital. There are many striking examples of it. PV projects in Denmark, not the sunniest place on the plane, were awarded for LCOEs as low as €17/MWh in 2016. Even though modules were much more expensive back then - thus increasing the total capex investment - the winning LCOE was exceptionally low. How was this record-low price offer then possible? Cost of financing capital and profit expectation, the reduced payment risk from a strong creditworthy buyer, are the financial drivers behind it.

Before going through the effects of financing costs, we explain how nameplate DC and AC system capacity ratings affect the project economics. We take this detour to show how, perhaps counterintuitively, Project IRR and LCOE do not always rise and fall together.

DC to AC Oversizing

Modules are rated in DC watt-peak (W_p) capacity at the STC reference. Inverters are rated in kVA, fixing a PV system AC W_p

capacity – the peak active power P output of the whole system when the total inverter kVA is utilized. In actual operating conditions, we've mentioned a few times, the STC are rarely replicated. Most of the time, for a 1:1 DC: AC W_p capacity ratio, the PV array is operating at 20-40% lower power output than the system AC W_p capacity (the inverter kVA max). Annual power attenuation increases the underutilized inverter capacity with time. Therefore, without adding any cost for additional inverter capacity, the DC W_p can be oversized to increase energy yield. DC W_p is increased for the incremental cost of the modules, mounting, and DC cables only - the rest of the system specifications and costs are largely unchanged. Only ~40-50% of the total system $/Wp applies to the oversized DC capacity, while the energy yield of Wh/W_p is increased for the whole system output. Revenue scales by energy yield so incremental revenue gain vs incremental cost limited to the modules, mounting, and cables only, improves the system economics. Additional kWh from the energy generating DC side is extracted without increasing the cost of the AC side (the AC subsystem is merely a conduit to the grid, with a slack in it for 1:1 DC:AC nominal rating). Energy yield and revenue increase disproportionately to the increment in total system cost by closing the inverter underutilization gap. Project IRR is increased and LCOE is reduced. There is a simultaneous benefit to the seller and buyer in reducing the underutilized capacity. This underutilization is an artifact of how PV systems are nominally rated in arbitrary technical definitions.

There are two points to emphasize about the planning of this mutual benefit to seller and buyer. The first is that engineering conventions or jargon of definitions, if not properly understood, can be needlessly detrimental to both the seller and buyer. Science and technology are replete with obfuscating or misleading jargon. The second is more profound. Mutual benefit has been created by tapping the underutilized yield of solar energy. Greater energy delivery, the link between the Project IRR (the seller motive is to maximise it) and LCOE (buyer wants to minimize it), is the enabler. Technology innovation can keep improving PV system economics by enhancing the energy delivery for rated nameplate capacities - increasing the energy delivery is not just about boosting power conversion efficiency but smarter utilization of the project conditions. Project IRR and LCOE dynamics are governed by engineering design and project planning - since there is a large gap between the raw available energy on tap (continuously received photons) and that being converted to useful purpose.

The addition of modules is constrained by available space. At increasing levels of DC oversizing, when the output exceeds the peak rated capacity of the inverters, the inverters reduce the DC power intake (by MPPT adjustment). There is a trade-off between peak power clipping losses and underutilization of STC rated capacity. For sites with low irradiance, the trade-off is in favour of even higher DC:AC oversizing ratios (less frequency of high irradiance clipping). Greater DC:AC oversizing is also well suited to PV systems installed with an East-West orientation (instead of the typical N-S) - especially beneficial for systems installed on a flat roof.

DC: AC oversizing ratios up to 1.3: 1 are common, in rare cases going as far as 2:1. On pure financial considerations, the optimal ratio for the project conditions is most often around 1.5: 1. Feasible oversizing has to be confirmed with the inverter manufacturer. For increasing DC: AC ratios, the inverter switching frequency and clipping become a technical constraint due to overheating of its electronic circuitry. Typically, inverter manufacturers recommend 1.35: 1 for safe limit.[4] Heat dissipation is an upper limit constraint on oversizing. An additional constraint might be imposed by the grid operator. Underutilized total apparent power AP capacity is reserved as emergency reactive power support ($AP^2 = P^2 + Q^2$ on the P-Q operational space limit, Ch6), for the grid operator to use at even full active power capacity. If the grid utility only fixes the AC W_p capacity of the PV system grid connection, DC oversizing is almost always financially beneficial.

Cost of Financing Capital

Project developers don't fund energy generation infrastructure projects out of their own pocket. They can access third-party finance at a lower rate than the Project IRR. Their return on equity IRR-on-equity, aka Levered IRR by debt to equity loading, is enhanced by the difference between the Project IRR and the Cost of Capital (CC) from a third-party financier.

We illustrate the cost of financing effect, in a hypothetical competition between two solar electricity sellers A and B. To focus on the financing effects, both sellers have identical systems with the same capex and O&M costs, and power performance profiles except for one characteristic that we will vary: the annual power output attenuation rate. AEY1 is the same for both A and B, but the energy output will slide differently for subsequent years.

Seller A secures a financier who provides them with \$3.2 million to be repaid back in equal principal repayment instalments of \$640.000 over five years, starting from the end of one year for the construction and commissioning, with a fixed annual coupon of 4% on the total loan i.e., \$128,000 per year on top. In return for the total upfront financing cost, including project planning and development, A pays \$768,000 to financier at the end of year-zero and each of the first four operational years (from the project revenues: product of annual energy yield and fixed tariff).

Total repayment over five years amounts to \$3.84 million. Cost of this financing capital "CC" - the borrowing cost of seller A = the compounded annual rate of return to financier - comes to 6.4%.

This debt leveraged IRR, "Levered IRR", amplifies the IRR-on-equity when Project IRR > CC. For CC < Project IRR, the IRR-on-equity worsens. The lower bound of LCOE for seller A can is thus practically set by their CC access.

Levered IRR vs LCOE options for seller A become:

CC 6.4% and tenor 5y			
LCOE $/MWh	Levered IRR	Project IRR	Gain
10	-8.7%	-6.6%	-2.1%
20	4.5%	5.0%	-0.4%
30	13.1%	11.0%	2.1%
40	22.5%	15.9%	6.6%
50	34.1%	20.3%	13.7%
60	48.6%	24.5%	24.1%
70	66.1%	28.4%	37.7%
80	86.1%	32.1%	54.0%

The Levered IRR boost depends on a combination of the (Project IRR – CC) spread and the tenor of 5 years for the financing repayment. Tenor and asset leasing or financing period are interchangeable terms in this context. The larger the (Project IRR – CC) spread or the longer the tenor, the greater the Levered IRR uplift. A practical, commercial manifestation of our financing example would be asset finance for the total upfront cost. Countless debt-equity financing combinations with different debt-equity ratios, debt maturities and sculpted principal repayments are possible in securing third-party finance - in each case the financial leverage is driven by the (Project IRR – CC) spread and the duration of the spread in the debt tenor. Different debt-equity combinations, with corresponding weighted costs of capital, can be translated to an equivalent blended finance CC and its tenor.

Seller B enters the competition, offering a similar PV system to the grid utility with identical performance, capex and running costs, but guaranteeing a 2x higher annual power attenuation rate from year 2 to 30 of 0.6%. Might be because they are more cautious in guaranteeing the minimum energy for the contract to avoid a penalty shortfall, lack the OM skills, or their module has been tested to have a higher projected power attenuation rate.

LCOE vs Project IRR for A and B

	Seller B	Seller A	
LCOE $/MWh	Project IRR	Project IRR	A -B
10	-9.0%	-6.6%	2.40%
20	4.5%	5.0%	0.50%
30	10.6%	11.0%	0.39%
40	15.6%	15.9%	0.34%
50	20.0%	20.3%	0.32%
60	24.2%	24.5%	0.30%
70	28.1%	28.4%	0.29%
80	31.8%	32.1%	0.28%

Unsurprisingly, seller B manages a lower Project IRR than A for each LCOE value to offer. B's handicap increases for the lower range of LCOE options - prices more attractive to the buyer. Keep in mind the LCOE is the selling price for the seller and buying price for the buyer i.e. the per-kWh or per-MWh electricity purchasing tariff. Too low a tariff and the PPA duration is restricted by annual revenues falling below annual OM cost.

Seller B finds a financier who provides the total upfront cost of \$3.2 million for annual principal repayment instalments of \$320,000 over ten years plus a fixed annual coupon of \$120,000 at the end of each of the ten years - an annual repayment of \$440,000 for ten years.

The cost of this capital, the compounded rate of return to the third-party financier, is about the same as for A: 6.3% for B vs 6.4% for A. Longer-tenor financing gives B an overriding arithmetic advantage.

The tables are turned on A by B for prices just above \$25.9/MWh.

Levered IRR vs LCOE for A and B (similar CC, longer tenor)

	Seller B 6.3%. 10y	Seller A 6.4%. 5y	
LCOE \$/MWh	Levered IRR	Levered IRR	A -B
10	<<0%	-8.7%	
20	2.3%	4.5%	2.3%
30	14.6%	13.1%	-1.6%
40	39.6%	22.5%	-17.2%
50	80.6%	34.1%	-46.5%
60	123.9%	48.6%	-75.3%
70	200.7%	66.1%	-134.6%
80	210.5%	86.1%	-124.4%

The Project IRR and LCOE are not just affected by the financial cost of capital but how the principal repayment is structured – a technique known as debt sculpting. What if the lender to B demanded a higher coupon repayment for the credit risk of the longer tenor? B agrees to pay a higher annual coupon of $180,000 a year for the ten years tenor, a total of $500,000 a year including the principal repayment, to their financer for a CC that comes to 9.1%. Seller B still achieves higher IRRs for LCOE offers above $33.1/MWh (while the third-party financier, the lender, also makes a higher return of 9.1%).

Extending the tenor works for a seller as long as their Project IRR is higher than the increased CC for the extended period, i.e., Project IRR versus CC measured over the additional five years only. There is a minimum LCOE floor, given declining revenues, for this extended tenor condition. B can still beat A in offering a lower LCOE, above this minimum, by securing a longer tenor.

Levered IRR vs LCOE for A and B (higher CC, longer tenor)

	Seller B 9.1%. 10y	Seller A 6.4%. 5y	
LCOE $/MWh	Levered IRR	Levered IRR	A -B
10	<<0%	-8.7%	NA
20	0.8%	4.5%	3.8%
30	10.8%	13.1%	2.2%
40	27.5%	22.5%	-5.0%
50	59.1%	34.1%	-25.0%
60	96.8%	48.6%	-48.2%
70	172.0%	66.1%	-105.9%
80	196.8%	86.1%	-110.7%

If both sellers have to achieve an IRR hurdle of at least 25% (perhaps a requirement of their shareholders), B can achieve it by bidding a lower price of $38.5/MWh while A has to reach $42.2/MWh. In the competitive bidding for such projects, $4/MWh is a wide margin to clinch the project win.

The difference in the cost of capital structuring has put B in pole position in terms of LCOE $/MWh, even though A is guaranteeing a higher power performance system with 2x better annual power attenuation from the second year.

Seller A bids $50/MWh while B bids $49/MWh.

Is selecting B the right choice for the buyer?

Is the lower LCOE the right choice?

It is well known that access to cheaper financing often determines what tariffs can be offered and who commercially wins projects. Low $/MWh LCOE projects are marketed for international acclaim. Seller B seems a clear winner in offering the lower LCOE tariff by a significant margin of $1/MWh - enough to win a lowest LCOE tender (buyers are often legally obliged to select the lowest LCOE, even if it is the thinnest of margins).

Consider two realistic scenarios for a grid utility buyer: their consumer billing price is at least $100/MWh for the project duration, or the next cheapest source of electricity they can access is $100/MWh i.e. the cost of electricity being replaced (the avoided cost of electricity).

Although the B tariff is lower, they are also offering less electricity to the grid utility. Seller A is offering 24.8GWh more than seller B over a 30-year period. Selecting B delivers greater initial savings, a total of $69,037 over the first seven years, but it is A who delivers greater cumulative savings over the whole 30-year system lifetime.

Given a consumer billing price or avoided tariff cost of $100/MWh (kept constant for the project duration, for simplicity of comparison), **seller A would deliver** extra income or savings respectively of $719,643 over thirty years for **an IRR gain of 14.4% to the buyer** (an NPV deficit of $207,808 for the utility at a financial discount rate of 5%).

Accepting B is a significantly lower-IRR option for the buyer, under these conditions. Seller A's higher 30-year IRR for the buyer emerges as the correct choice, despite a higher PPA tariff offer.

Simply choosing the lowest LCOE on offer might not be the most advantageous decision for the grid utility. Instead, the grid utility should seek to maximise the net present value of their associated income or levelized avoided cost of electricity:

$\sum_{n=1}^{n=30}$ (Consumer Billing Price – LCOE)$_n$ • $E_n/(1+d)^n$, or

$\sum_{n=1}^{n=30}$ (Avoided Electricity Price – LCOE)$_n$ • $E_n/(1+d)^n$,

where "d" is a discount rate for the buyer's time value of money.

The flaw with LCOE for a figure-of-merit is that the electricity sellers are offering a vector combination of cost/kWh and E_n (the value of the offer is defined by two dimensions of cost/kWh and E_n). LCOE treats the two-dimensional vector offers in a single dimension for a suboptimal comparison, losing information about the value of electricity (the differential quantity value of electricity delivered). A vector constraint boundary for a two-dimensional vector figure-of-merit is required that does not curtail the performance quality, i.e., production volume, of the competing offers. The consumer billing price or avoided cost of electricity is an appropriate vector boundary constraint for dynamic optimization, avoiding a static performance threshold. The techno-economic value formulas proposed above are volume-weighted objective functions. These formulas generate the total objective value of the electricity magnitude. The objective value is constrained by the price at which electricity can be sold or the avoided cost of buying more expensive electricity. Consumer billing price is preferable for the constraint.

Tenders that award winning bids solely on the basis of the lowest LCOE, even with thresholds for technical criteria, can often be suboptimal for the grid utility – possibly by millions of dollars. The lost financial value, of a short-sighted LCOE decision, scales by the size of the project.

We deliberately chose the year 2 to year 30 power attenuation rate for the sole distinguishing factor between the two bidders – in order to highlight the critical dependence of financial models on actual versus projected simulations of system project performance. The buyer must ensure the seller has planned adequate quality assurance measures, such as the field-testing techniques described in previous chapters, to verify the actual power attenuation. In addition, the legal agreements must contain provisions to enforce actual performance meets the projected expectation. Diligent project planning thus requires careful alignment of financial modelling, technical and legal measures.

Performance Guarantee

Financial planning is not only about the tariff numbers but also risk management - the seller and buyer have different perspectives. Solar electricity sellers generally do not take liability for matters outside their control that may affect the system kWh output, e.g., the actual solar irradiance and temperature at the site versus what was simulated. In short, sellers do not take weather risk or liability for equipment under the control of the grid operator. Hence, the importance of the industry standard PR definition - an adjustment for the actual irradiance and temperature received versus the STC. The simulated energy generation schedule submitted to the utility for the grid impact study has a guaranteed PR column - typically calculated according to the PR_{STC} defined in the previous chapter. A standard legal arrangement is that the seller is obliged to guarantee this PR_{STC}, adjusted for deviations between the simulated and actual irradiance, temperature conditions, and any other agreed loss factors (e.g., the assumed grid unavailability). The minimum guaranteed energy schedule for the PPA becomes a minimum guaranteed PR schedule.

The guaranteed PR of the seller is given by:

Guaranteed PR_{STC} = Measured PR_{STC} • Corrective Factor for Simulated vs. Measured Loss Factor.

For example, the Corrective Factor for temperature =

1 + ɣ • (measured module temperature – simulated module temperature), where ɣ is the temperature coefficient of power loss from STC for the module (Ch8.). For commercial modules, ɣ is negative - expressed as a percentage of power loss from the maximum power point per unit temperature.

The seller and buyer should discuss appropriate adjustments for each loss factor in the energy simulation that is not under the complete control of the PV power plant operator. There is a modelling uncertainty in the simulated loss factors, as well as measurement uncertainty, to account for in the comparison of the calculated loss factors. Even with perfect measurement sensors, there will be some residual uncertainty.

The display of simulated results for both P50 and P90 exceedance probabilities is standard practice. P50 or P90 means there is a 50% or 90% probability of the shown energy generation being met or exceeded. It depends on the accuracy of the simulated loss factors, so the corrective factors are essential.

The grid utility buyer should not only rely on PR measurement to validate the system has been properly installed, according to the agreed design. For example, the orientation of the array has to be independently confirmed - the PR formula does not account for a discrepancy in the planned GTI.

The module annual attenuation rate has a different value for the first year of operation, than the following years, so a performance judgement on the basis of the first year's results is insufficient. The financial model for the projects rests upon technical assumptions that have to be field-tested and verified.

Legal Agreements

The financial closure of the project requires a basic set of documents: the PPA, GCA, land lease, financing agreement, and an EPC (engineering, procurement and construction) contract if the project developer is outsourcing the whole project works. The project works contract contains payment milestones, based on quality assurance tests, for mechanical and electrical works. In addition, the project developer must have all their supplier and subcontractor contracts in place. The most important ingredient of the supplier purchasing contracts is the quality control measures, in particular meticulous FAT validation.

The PPA is an elaborate document with many topics warranting careful attention. As with the GCA, it is critical to have continuous technical oversight of the drafting till its conclusion. A comprehensive set of legal provisions is necessary to ensure the enforcement of technical obligations, careful precautions, quality control measures and remedies. We discussed how maintaining the grid connection for smooth power flow depends on both sides. High-importance PPA terms include:

- Termination causes, termination value assessment, and buyout clauses
- Precise and unambiguous tariff formulation
- Performance shortfall and time delay damages
- Change orders
- Change of law
- Compensation for interruptions caused by the grid utility
- Precise technical definitions, e.g. of the guaranteed PR
- Provision for independent expert, technical adjudication
- Technical audit records including the Provisional and Final Acceptance Testing for commissioning

Most importantly, apart from being comprehensive, the terms should be clearly spelled out. Ambiguity is the bane of technical legal contracts. The project developer submits a detailed final design with the PPA contract. Technical appendices must be completed in full detail as an integral part of the contract. Agreements relying on catch-all clauses are a bad sign of lazy negotiation. Despite the intention, it inevitably leaves unintended loopholes to cause trouble later.

Tightly negotiated legal terms are not a zero-sum game for the sellers and buyers. Secure payment terms for the seller can also benefit the buyer. Reducing payment risk enables the seller to arrange longer-tenor financing capital, driving down the LCOE it can offer to the buyer. The seller receives a higher IRR, the financier higher returns on their capital, and the buyer a lower per-kWh electricity purchasing price.

Financial Exits for Project Developers

Project developers seek financial exits from their solar assets once they have secured operational cash flows i.e., the PV system is generating energy and proportional revenue as expected. Solar asset financials do not have any fuel price volatility. Provided the PV system does not have any latent defects and chronic underperformance issues, it represents a very safe investment with steady, predictable cash flows. The LCOEs of fuel-based electricity generation assets are always speculative because no one can lock in fuel prices for the duration of the project lifetime (hedging of fuel prices is only available for limited periods). The solar resource is highly predictable over the long-run, with interannual standard deviations of ~3-5%, perhaps 10% at the most for coastal areas with volatile weather. The cumulative standard deviation of projected cash flows is reduced over the project duration, of ten to thirty years, to a fraction of one percent.[5] For fuel-based generators, operational costs of the LCOE numerator are not fixed in the same way as for PV projects. The standard deviation of the projected cash flows of the fuel-based generators increases over the project duration.

Therefore, there is a substantial market of buyers for solar assets who want to procure safe long-term cash flows with an underlying physical asset, particularly for PPAs with creditworthy counterparties. Savvy project developers forward sell the operational cash flows of the solar assets, prior to construction. This strategy helps in obtaining lower cost of capital construction finance for a more competitive LCOE. A lower LCOE helps in winning the project PPA. The project developer can plan their LCOE offer to the PPA counterparty, knowing the yield they have arranged for selling the operational cash flows. Forward selling of the operational cash flows for a yield X% < Levered IRR Y%, with a net present value (NPV) profit given by

NPV Profit = (PV of cash flows discounted at X%) minus (PV of cashflows discounted at the Levered IRR Y%).

Solar asset owners may sell subscriptions in the operational cash flows to retail investors, e.g., structured as a financial bond (sometimes called green- or eco-bonds). The cash flows might be packaged as an asset-backed security for financial trading. “Yield Cos”, publicly listed companies offering shares in the financial yields of solar assets (often packaged with other renewables in a diversified portfolio), is another variation. The stock price of the Yield Co is meant to reflect the discounted net present value of the cash flows of its underlying asset portfolio. All of these financial schemes follow the same mathematical principle, with the retail investors offered an asset cash flow yield of X% < Levered IRR Y% (the Levered IRR is the IRR-on-equity of the project developer).

Community Solar

The community solar business model is a different concept from offering retail investors a share in the financial yield of a solar asset. Community solar offers people a share in the electricity production of a solar asset, to offset their electricity consumption from the grid. The equivalent revenue from an electricity production share will be higher than the financial yield offered by a project developer for bond-like subscriptions.

A community solar participant can pay for their share in the financing of a solar asset, through their electricity bills. A proportional share of the solar asset energy yield is credited to reduce their electricity consumption. The community ultimately owns the solar asset. Each community member has an ownership share in the asset, paid for through their electricity bills, with equal voting rights for its management and governance. It is the electric cooperative business model, applied to solar assets.

Crowdfunding or crowdsourcing is different from solar bonds, yield cos, and strictly speaking, also from community solar. It is the more general practice of raising funds from numerous people, typically through a Web platform, to finance a project. The funds might be donations, rather than investments in shares.

Due Diligence of Solar Assets

Solar asset buyers, typically institutional finance e.g., pension funds or insurance companies, are proficient in assessing credit risks, the creditworthiness of the PPA counterparty, and legal arrangements.

In-depth technical due diligence is more challenging, e.g. the fine detail of the grid connection agreement, or the technical assumptions behind the legal provisions and financial projections. Technical scrutiny of the asset performance is a layered process. Defects, degradation mechanisms, and latent performance issues can unintentionally or deliberately be masked in the headline reported data and broad performance metrics. If the asset is being bought early, defects and degradation mechanisms might still be growing to become a permanent system handicap - though not yet discernible in the electrical data. For example, a flawed system mounting structure might not have experienced severe weather conditions exposing latent weaknesses. PR measurements do not reveal leading indicators of system problems. When PR is not being measured properly, the system performance results might be inflated by above-average weather conditions.

Technical due diligence must check the history of how the project was constructed, starting from the traceability of the procurement of the system components. A professional project developer will have prepared a detailed method statement for the project works, including the FAT, Site Acceptance Tests, quality assurance tests for milestones of the civil, mechanical, and electrical works, Provisional Acceptance Tests for commissioning (aka the commercial operation date), a Final Acceptance Test after a minimum of two years (the annual power attenuation is different for the first and second year), and post-commissioning field-tests for operations and maintenance.

We described several different types of field-testing techniques to investigate a wide variety of possible defects and degradation mechanisms in the PV modules. Field tests and operational monitoring should cover all the system components: modules, inverters, mounting, cables, transformer, and switchgear. Relying only on the FAT records is insufficient. Frequently, if not most often, the defects are introduced during installation. A comprehensive project quality assurance plan would have monitored the work of each installer e.g., EL imaging inspections of sampled modules per installation team, checking the cable connections were made with the proper tools, or the cable insulation was not damaged. Solar assets without a detailed technical audit trail are risky propositions.

Studies indicate that PV systems can actually degrade at a multiple of the initially projected rate, with corrective actions required for the failures of different components. A survey of 21GW utility-scale projects in the US found average annual power attenuation rates of 1.3% for the installed systems.[6] A study of 3.3 million modules in the UK found ~37% had thermal defects, including multiple hotspots and heated substrings.[7] A US renewables firm conducted EL imaging on 300,000 modules across 150 projects in 16 countries (including the US, Canada, Australia, and Western Europe) - finding 80% of the projects had defective modules.[8] Bypass diodes and backsheets are often overlooked in the module quality control. Other investigations have focused on failures in inverters[9], the mounting[10], and the cables[11]. Unattended defects can lead to electrical fires and safety hazards. These premonitory results are for markets where generally more attention is paid to quality control and the installers are better trained. Defects can emerge after 5-10 years, hence field-testing for early warning signs is crucial. Comprehensive field-test regimes help distinguish latent manufacturing defects from installer mistakes, for warranty enforcement. Conversely, diligently procured, installed, and maintained PV systems exceed power performance expectations, with a longer productive lifetime than thirty years.

When there is no detailed technical documentation to audit, the solar asset buyer is well advised to carry out their own testing protocols. Smart I-V scans by the inverters quickly reveal anomalies in the strings' performance. Portable EL imaging can be conducted on individual modules or entire strings. Drones are the fastest method to analyse in-situ system performance in different operational conditions. All the more reason to conduct these tests, if the solar asset owner has been negligent about it. Adequate provision of spare parts is a basic checklist item. So is confirming the warranties and insurance policies in place. Confident financial projections of the project cash flows is impossible without a comprehensive testing record in actual operating conditions.

Rooftop Customers

Solar rooftops essentially consist of two types of projects: one for self-consumption only, and in the other, the rooftop owner is allowed to export excess electricity to the grid for some compensation. The compensation schemes come under several different names, with a variety of jargon and bureaucratic red tape. Solar rooftop owners face a substantial disadvantage in their smaller procurement orders. They cannot afford FATs, technical advisors, or negotiating contracts with detailed quality assurance measures and performance obligations. They are largely at the mercy of the installers or the rules prescribed by the electricity regulatory authority. Aggregated procurement by communities can address some of these issues.

For the self-consumption customer, the first planning step is sizing the rooftop system not just for the available space but their expected load curve – fitting the expected system output profile to it, to avoid installing excess capacity. The load curve is an hourly estimate of the various demand loads. Closely aligning the loads to the PV output profile is particularly beneficial for low-income households, reducing their bills by timing the use of appliances (e.g. an electrical cooking stove) for peak solar production hours.

Asking the installer for a system energy simulation is a basic requirement - this technical data is easy to produce for any location. The installer should also provide flash tests and EL images of the modules they have purchased and are deploying for the customer. Confirm module serial numbers match the technical documentation. Find out who to contact for warranty enforcement. All modern inverters come with zero reverse power control capability, typically with a data logger for the purpose. Inverter manufacturers provide recommended circuit configurations for the data logger. A circuit-breaker, or protective relay, between the PV system and LV distribution board of the premises is essential. Inquire about the inverter safety features (Ch4), e.g., the Arc Fault Current Interrupter. The placement of the inverter is important, preferably in a shaded place. Learn how to safely de-energize the system – know where to find the emergency disconnect switches. The provision of a fireman's switch, aka a PV cut-off switch, allows DC power to be safely and quickly cut off. Have thick gloves and shoes available for inspections, emergency action. Solar panels are high current devices, so leakage currents can be fatal. Make sure the installer conducts insulation resistance tests, among a complete checklist of commissioning tests for handover.

The rooftop owner can validate the string lengths designed by the installer, using the equations given earlier in this chapter for maximum and minimum string lengths. Confirm that the strings are connected as shown in the system design. Make sure the

installer double-checks there are no loose connections. No dirt is stuck in the connections or moisture ingress occurs to corrode metal contacts, when the connections are being made (it can easily happen if the installer is not wearing gloves, and has dirty or clammy hands).
Careful attention to the mounting structures is critical. Ask for evidence of scientifically modelled and tested designs. The installer may downplay the importance of the mounting structure but it can be the cause of harmful, if not lethal, accidents. It takes only one sub-standard part to compromise the whole mounting structure. Panels should not be fastened in a manner that induces any stresses. Adequate grounding to earth is a basic necessity. Safe distances should be kept from the edges of the roof ~0.5-1m. Adequate space is also necessary for maintenance access and emergency work. Densely packing the panels on the roof, so that there is no room for inspection, is highly inadvisable - leave about ~0.2-0.4m minimum between strings, 0.2m only if microinverters are being used, with greater caution for tilted roofs.
The rooftop owners who are installing systems with grid export permission, should have the benefit of technical guidance from the distribution grid company: considering voltage regulation, transformer capacity, and ampacity of the grid power evacuation. It is up to the distribution grid company or electricity regulatory authority to specify the operational parameters for the grid connection. All modern inverters can act as voltage support

resources for the grid, providing reactive power programmed for voltage droop control, maintaining power factor, Q(V), Q(PF), P and Q dynamic controls, according to the grid rules. The inverters can also provide Q at night,[12] by drawing a small amount of power from the grid. After a steep slide in battery energy storage prices, hybrid inverters are an increasingly attractive option. In more advanced economies, or rather economies with more savvy grid operators, solar rooftop owners are participating in virtual power plants. Distributed energy resources (DER) are an increasingly powerful asset shaping the modern grid– a future of more economical and versatile, decentralized electric power.

The reader might have noticed that rooftop solar projects involve less hardware outlay than ground-mount, due to the use of existing LV/MV distribution boards and transformers. So why do solar rooftops require a higher LCOE tariff? There is a handicap in procurement scale, but a bigger reason is cost of financing capital. An individual solar rooftop owner does not have access to the same type of third-party financing for a higher levered IRR, as a professional project developer for a large ground-mounted project. If individual rooftop customers have no choice but to borrow the upfront capital at higher rates than their system project IRR, it increases their LCOE. Aggregated solar rooftop projects, with a low cost of financing capital, can achieve the highest financial returns - depending on customer acquisition costs for the aggregation.

A government or business service that helps individual rooftop owners come together for aggregated procurement has both technical and financial benefits. Joint procurement shares the financial overheads of quality-controlled procurement to reduce technical risk. It also helps the rooftop owners obtain favourable financing terms to reduce the LCOE of their installations. A lower rooftop LCOE means the government spends less on financial incentives for this distributed energy generation.

Free Solar

Marketing schemes dubbed as "free solar" are aimed at rooftop customers. These schemes can be good or bad, but as the reader might guess are never free. The "free" might refer to no upfront cost for the customer. The installer takes care of the capital expenditure for deploying the PV system on the customer's rooftop, and charges a $/kWh fee for the kWh savings from the reduced electricity bills of the customer, i.e., the kWh substituted by PV generation (or it could be in return for the kWh bought by the grid operator). Essentially, the installer is selling solar electricity as a service, or equivalently a performance-based lease of the PV system to the customer. Given fair pricing of a fixed service charge, or lease payment, this business model has significant merits for the customer. The customer does not have to worry about maintenance, and to a certain extent, feels comfortable that the installer will strive for high performance output (since the payment to the installer is in $/kWh for the actual kWh delivery). There is technical risk mitigation in this commercial engagement for the customer. It also helps customers who cannot afford the upfront capital expenditure themselves (though PV system cap ex has sharply dropped), to access solar power. The issues for careful attention are fair pricing terms, and termination clauses without any encumbrance for the customer if the system underperforms.

In another type of "free solar" scheme, the installer may take care of the upfront capital expenditure and offer daytime consumption from the PV system, in return for grid export tariffs. Effectively, the installer is renting the customer's roof in return for a quota of daytime power consumption, to generate income from solar power export revenues to the grid. In such a situation, the customer has to be wary of how the system is configured, i.e., whether they will receive sufficient self-consumption power, and the export revenues they are forgoing for themselves is worth it - the installer might be making a killing from a generous export tariffs scheme that the customer could have availed.

Role of Distribution Grid Utilities

What's the incentive for distribution grid companies in offering export tariffs, or billing savings, to consumers for solar rooftop installations i.e., in the creation of electricity prosumers (producers and consumers)? A cynical view would be that the distribution grid company can entice consumers to invest in delivering solar electricity to the grid, and then roll back the compensation for it. The greater the electricity their consumers inject in the grid, the less the distribution grid company has to source from elsewhere, and the solar rooftop prosumers can only sell their excess solar power capacity to them.

DERs, e.g. solar rooftops, save on capital intensive and time-consuming grid expansion, e.g. connecting to a distant power plant for demand growth – costs that have to be shared by all grid users, the economy as a whole. A more enlightened view would be that decentralized resources make the grid stronger, more efficient, economical, and resilient for everyone[13] - provided the DER technology is properly utilized for grid integration.

DERs introduce greater competition in the market, forcing other generation sources to compete at lower prices. Lower electricity prices boost productivity across the board, making the economy more globally competitive.

Electricity is a necessary precursor for economic growth. It is a basic, pervasive input in economic productivity. Solar electricity, unlike fuel-based generation, is deflationary. Solar rooftops are an antidote for electricity inflation. Distribution utilities have to assist in the reduction of electricity inflation, enabling rather than inhibiting the rapid growth of DERs for a wider cross-section of society. For example, by offering community solar plans to consumers who do not have adequate rooftop space of their own.

Distribution utilities are typically not generators - vertically integrated utilities were broken up, so that there is greater competition in the electricity generation marketplace, and consumers have more choice. If the distribution utility is in the generation business as well, it can offer its own DER solutions to consumers, but competitively without a monopoly. Collusion between the distribution grids and legacy power plants to inhibit DERs would defeat this purpose. Organizational inertia and legacy relationships have to be overcome for forward-looking infrastructure modernization. Demand suppression, forcing consumers to either buy more expensive electricity from legacy sources or curtail their demand, stifles the growth of the entire economy.

DERs, in particular solar rooftops, offer new capabilities to modernize the grid.[14] The existing grid, a public investment, is used more efficiently. There are lower physical power losses, the shorter the distance between the generation source and its consumption load. Reduced power losses benefit everyone. Distributed solar rooftops, we shall see, can be configured to flexibly support the grid, e.g. replacing old, static, and expensive technologies to dynamically compensate for voltage drops – a major cause of power losses and grid instability. DERs provide a buffer against load-shedding in weak grids. Virtual power plants are viable for greater grid resilience and energy security.

DERs, in particular geographically dispersed PV systems, located closer to the point of consumption, alleviate the problem of grid congestion. Reliance on distant power plants is subject to this risk. Grid congestion occurs when the transmission lines cannot support the power flow for reasons of thermal, voltage, or phase angle stability, typically caused by sudden changes or re-routing because of supply outages. Long distance transmission lines, and concentrated power generation sources, are more vulnerable to grid congestion.

Power flow, P_{flow}, across the transmission and distribution grid lines is given by: $P_{flow} = (V_{source} \cdot V_{load} \cdot \sin \delta)/X$.

V_{source}: voltage at the generator sending end

V_{load}: voltage at the receiving end of the load

δ:phase angle difference between the voltage phasors at the sending and receiving ends

X: the line reactance

For different generators trying to send power through the same transmission line(s) the total power flow P_{flow} must satisfy the condition $\sum_i P_{flow,i} \leq Pmax$. The maximum power P_{max} limits for the lines are determined by their ampacity (thermal limit of conductor current capacity), voltage constraints, and the phase angle δ stability. Local PV eases these limits: creates room for spare conductor ampacity in long-distance grid transmission lines, and increases voltage support at the load receiving end to reduce line power losses and phase angle stress.

For a given power demand, the reactance X increases with distance, placing more stress on δ to maintain the required power flow. The P_{flow} from distant power generation sources has to be increased to meet the load power demand, causing grid congestion in overstressed, shared grid lines with maxed out ampacity. To pre-empt grid congestion, the alternative solution is costly and time-consuming expansion of transmission lines for greater redundancy in power flow paths i.e. if one line gets too stressed, the power flow can be rerouted. Such grid expansion costs are ultimately borne by all consumers.

Enlightened distribution utilities ought to embrace DERs, for the development of a modern grid that democratically empowers people in ways that was not possible before. The utilities have privileged licenses to distribute electricity, for the sake of economic progress. Distribution utility management has to be judged on how effectively they function in this context. Understanding optimal grid integration strategies for PV, including solar rooftops, is essential to their raison d'etre.

--*--

Ch9 references, footnotes, and supplementary reading

1. Geotechnical Analysis and PV Foundation Design, Bob Donaldson and David Brearley https://definitivesolar.webvent.tv/uploads/assets/4789/document/0001-SP8.3_FE_Donaldson.pdf
2. Publicly available samples of grid connection agreements: https://www.hutchinsonutilities.com/wp-content/uploads/2019/06/14_Interconnection-Agreement.pdf https://ppp.worldbank.org/sites/default/files/2021-09/FINAL_Zambia%20REFIT%20Grid%20Connection%20Agreement.pdf
   In practice, the project developer needs to negotiate technical details pertinent to their project, e.g. clearly and distinctly specifying the Point of Connection (POC) and Point of Common Coupling (PCC) – the POC and PCC might coincide, but not necessarily. It makes a difference for the joint implementation and maintenance of power quality standards.
   For rooftop solar, the POC would be the LV distribution board, while the PCC would be the secondary connection (on the customer side) of the LV/MV distribution grid transformer. POC is the system interface demarcating ownership of assets, while PCC is a physical coupling with the wider grid.
3. MC4 Y branch connector for solar: https://pantherrvproducts.com/pan33r/
4. Fronius highlighting a 1.5:1 DC:AC oversizing capability for their inverter: https://www.fronius.com/en-us/usa/solar-energy/installers-partners/info-center/technical-articles/the-hidden-power-of-inverters

5. Interannual solar variability for five different locations for 1 year, 5 years, 10 years – in a range of 0.2-0.5% for 10 years: https://kb.solargis.com/docs/solar-resource-variability
6. System-level performance and degradation of 21 GW-DC of utility-scale PV plants in the United States, 2020, Mark Bolinger et al. https://doi.org/10.1063/5.0004710
7. Field study on the severity of photovoltaic potential induced degradation,
Nature, 2022, Ghadeer Badran and Mahmoud Dhimish
https://www.nature.com/articles/s41598-022-26310-y
8. News article about module defects in international projects sampling:
https://www.pv-tech.org/massive-increase-in-solar-module-defects-poses-significant-risk-to-pv-industry-cea/
9. Failure Mode and Effect Analysis of Photovoltaic Inverter, IEEE Open Journal of Power Electronics, 2026, Buck F. Brown III et al.
DOI 10.1109/OJPEL.2026.368585
10. Failure of Solar Panel Aluminium Mounting Structures Subjected to Uplift Pressure: Effects of Foundation Defects, Solar, 2025, Furukawa et al.
https://doi.org/10.3390/solar5030033
11. Institute for Building Technology and Safety (USA): "our data shows that wire management problems occur on 38% of PV projects nationwide"
https://www.ibts.org/common-deficiencies-solar-pv-installations#
12. Using PV inverters at night can save grid costs, Energy Reports, 2022, Yamit Lavi and Jay Apt
"Using PV for voltage support at night saves millions in grid costs"
https://doi.org/10.1016/j.egyr.2022.05.004

13. Recent 2025 study of distributed PV Net Metering benefits to Delaware grid, commissioned by the state-utility Energize Delaware: https://www.energizedelaware.org/

"A large expansion of access to net metering could absolutely help reduce everybody's bill. That's what the study shows" – Delaware Sierra Club

14. Effects of distributed PV generation on California's distribution system Part I: Engineering Simulations, Solar Energy, 2016, M.A. Cohen and D.S. Callaway https://doi.org/10.1016/j.solener.2016.01.002

Ch10. Grid Integration and Flexible Dispatch

"Listening to both sides of a story will convince you that there is more to a story than both sides" – Frank Tyger

Decentralized power is generally a good thing. So is flexibility in system networks. We noted that all energy replenishment on the planet comes from the Sun as electromagnetic energy, and solar PV directly converts it to useful electric power. The conversion of the solar resource to electric power at any location does not affect what another place receives and can utilise at the same time – there is no zero-sum game involved or knock-on effects (in hydroelectric and even wind power, there is a downstream impact). PV systems therefore provide a method of raising the useful energy available for everyone.

Electricity consumers have become capable of producing their own power - so-called prosumers - at a cheaper cost than a large, centralized power plant. The materials have to be initially sourced but once set up, no consumable is required for every electric unit of kWh that is produced. Electricity generation cost is fixed and stable for the long run, to the extent the system is free of unexpected degradation.

There is no fuel input, so no fuel price volatility in the LCOE. Not only is solar electricity cheaper, but also fundamentally price stable. The LCOE can vary by access to money (the cost of financing capital), but that is a human artifact not a fundamental natural constraint. For PV systems with equal power conversion efficiency and financing costs, the LCOE will be lower for areas with higher insolation.

It may therefore seem sensible to concentrate PV systems within a grid jurisdiction in the areas of highest insolation, for the lowest LCOE per project. Such a strategy does not take full advantage of the grid, and the ubiquitous distribution of the solar resource (insolation varies by a maximum difference of ~3x across the globe, and much less regionally - it's trivial compared to the geographic variation of other natural resources). On the contrary, a lopsided concentration of PV on the grid can prove detrimental.

Electric power demand on the grid is constantly changing on different timescales. It is not technically feasible or economical to crank a single type of power source up and down to meet every demand fluctuation. A certain type of electric power source might be much more economical than another resource for a shorter, longer, or different time period. Over the long-run, demand load patterns are shaped by the mix of electric power sources made available on the grid, and related strategies like time-of-use billing.

Baseload and Merit Order

The term baseload power is often used rather misleadingly. Electric power delivery for any time period is not based upon a physical stack of materials, like bricks and mortar construction with a base foundation. Electric power flow from independent generators is seamless, provided there is no grid congestion. The default dispatch order is set by relative production cost, with the cheapest power source available for the time slot dispatched first to meet as much demand as it can, followed by the next according to the demand variation, and so on. Typically, the dispatch order for each hour is scheduled according to a day-ahead demand forecast. The dispatch sequence by lowest cost is called the merit order (the relevant cost is the marginal or operational dispatch cost).[1] In this sense, the cheapest power source for the given time slot is the baseload power.

Fuel-free electric power offers the lowest marginal cost of dispatch. PV and wind power have a fundamentally lower marginal production cost, not just because of no fuel input but also because there is no water input in the kWh generation. The term baseload can have a different interpretation: the baseline demand that is expected for a given time period - in this context, adequate baseload power resources backed by sufficient redundancy have to be scheduled for each period (with the most economical ready to be dispatched first).

The other key criterion for the power generation dispatch order is sourcing the generators that can provide a necessary system response to manage particular grid operational conditions. Generators with black-start capability may need to be dispatched first to set the grid frequency and voltage (e.g., after an outage), followed by the economic merit order of lowest marginal production cost. Scheduling of active power ramps for unexpected deviations from demand (real-time supply shortfall for peak demand), or emergency voltage and frequency regulation provisions, are assessed. PV grid integration, and the whole electric power portfolio, has to be planned with these considerations in mind: the merit order of lowest production cost, redundancy, availability, grid frequency and voltage stability, power fluctuation tolerance, and control for necessary system responses. Options like synchronous condensers, the resurgence of a 1950s technology, decouple grid frequency and voltage setting from the dispatch of active power. Stranded assets of thermal power plants, that are no longer economical in their original role, can be reconfigured to function as synchronous condensers.

Figure 9: Actual electricity generation visuals for Germany grid

CC BY 4.0 license – source: SMARD.de Bundesnetzagentur

August 31st, 2024 – hourly resolution data, y-axis is MWh

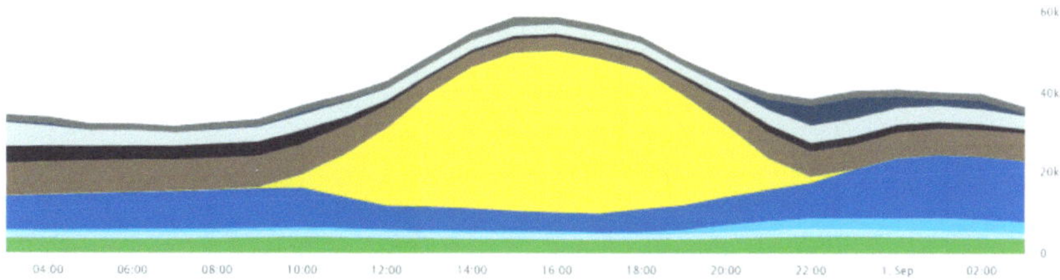

March 1st, 2025 – hourly resolution data

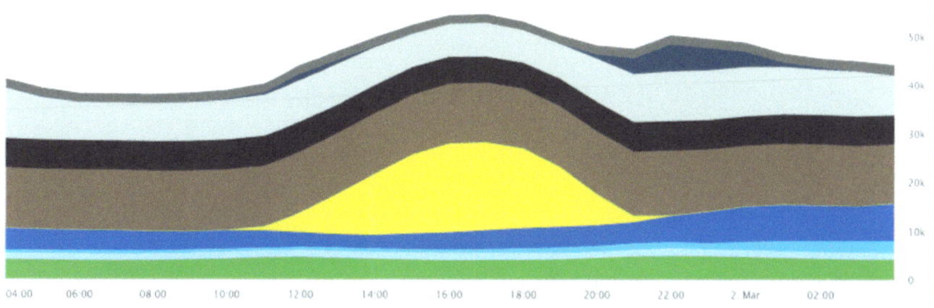

Dispatch merit order from bottom:

Green: **1. Biomass** (e.g. biogas from organic waste), Pale Blue: **2. Hydropower**, Light Blue: **3. Wind Offshore**, Dark Blue: **4. Wind Onshore**, Yellow: **5. Solar PV**, Brown: **6. Lignite**, Black: **7. Hard Coal**, Grey: **8. Fossil Gas**, very dark Green at the top: **9. Pumped Hydro Energy Storage**.

We have shown the merit order of dispatch for a summer and winter day in Germany. The merit order is affected by the types of power purchase agreements in place for the electricity generation source. For certain types of generators, priority dispatch might be required for technical and or legal reasons other than lowest marginal cost.

Spatial PV distribution

The spatial, or geographic, distribution of PV resources on the grid has to be planned in a system-wide context. During the daytime, when it is available, PV has a zero marginal dispatch cost, since there is no operational fuel cost. PV system availability, defined as the ratio of actual system operating time to the total time it should be online, is among the highest if not the highest in the electric power industry. Greater than 98-99% system availability is contractually stipulated for PV systems. The planned and unexpected downtime of the system is restricted to less than 1-2% (the time that the system should be switched on is defined by a threshold irradiance level ~30W/m^2 GHI for the inverter), subject to exceptional circumstances beyond the control of the system operator. The amount of power that the PV system can deliver depends on the weather conditions at the time, mainly the irradiance. PR measurement rates the system for its power delivery performance, subject to these weather conditions. System availability is factored into the PR calculation.

Extraterrestrial solar irradiance is largely determined by the relative astronomical positions of the Sun and Earth, in the shape of a trapezoidal or bell curve. Terrestrial solar irradiance on a clear day, known as clear-sky irradiance, approximates the ideal, deterministic bell curve, but with noise superimposed on it. By deterministic, we mean its temporal variation can be accurately predicted. Predictable variability is not so much a problem for electric power scheduling, as long as it is not that sharp and frequent. It is the unpredictable variability that increases the need and costs for backup power resources.

Clear-sky irradiance excludes clouds, but includes aerosols, water vapour columns, and atmospheric gases. Aerosols primarily and water vapour secondly cause intra-day and inter-day variations in the clear-sky irradiance. A conservative estimate for intra-day temporal variations from the ideal bell curve for any hour would be around +/- 5%. Sub-minute variability is generally insignificant in clear-sky conditions but unpredictable hourly or sub-hourly variation on the order of fifteen minutes deserves attention. Inter-day variation, even under clear-sky conditions, can rise to +/-

10% or exceptionally +/- 20% in areas heavily affected by aerosol pollution.[2,3,4,5]

The temporal variability of the solar irradiance between two different locations is quantified by a statistical correlation coefficient (SCC) that measures the strength and direction of a linear relationship between two variables. It ranges from – 1 to +1: -1 indicates the two variables move exactly opposite to each other, 0 that the variability of one variable is independent of the other, and +1 that they vary identically in the same direction. SCC is the covariance between two variables normalized on a scale of −1 to +1. The clear-sky irradiance variability (a measurement of variability in the GHI time-series for a location), decorrelates with distance – it is called spatial decorrelation or the decorrelation length. The spatial decorrelation starts to become significant after ~10-20km: SCC dropping to 0.5-0.7 at ~50km, 0.1-0.3 for ~200km, and 0-0.1 after 500-1000km.[6,7,8]

As the number of PV projects is increased over larger decorrelation lengths, variability in the aggregated PV output is reduced by a sheer statistical effect (the law of large numbers). The standard deviation (S.D.) of the aggregated PV output is related to that of an individual system as follows:

$$S.D\text{-aggregated} = S.D.\text{-individual} \cdot [(1 + (N - 1) \cdot SCC)/N]^{1/2},$$

where S.D. is the standard deviation of the irradiance-dependent PV production, N is the number of PV systems, and SCC is the statistical correlation coefficient between −1 and +1.

This formula assumes the N number of PV systems are roughly the same capacities. The more general equation for systems with differing capacities is given by

S.D.-aggregated = S.D.-individual • $((\sum_{i=1}^{N} \sum_{j=1}^{N} w_i \cdot w_j \cdot SCC_{i,j})/(\sum_{i=1}^{N} w_i)^2)^{1/2}$.

Where w_i, w_j are the capacity weights of the system, and $SCC_{i,j}$ is the correlation between the sites.

When all the PV systems with different capacities are completely uncorrelated it reduces to S.D.-aggregated = S.D.-individual • $(\sum_{j=1}^{N} w_i^2/(\sum_{j=1}^{N} w_i)^2)^{1/2}$.

When all the PV systems have roughly the same capacity and are completely uncorrelated, S.D.-aggregated = S.D.-individual/$N^{1/2}$

More realistically, there will be some spatial correlation but the aggregated variability can be reduced e.g.:

For a geographical dispersion radius > 200km, SCC ~ 0.2, it will take 16 sites of roughly the same capacity, to reduce S.D.-aggregated to +/-2.5% and +/-5% for the intra- and inter-day variability respectively (for S.D.-individual of +/-5% and +/-10% respectively). Increasing the number of sites has diminishing returns, so the spatial smoothing for SCC ~0.2 saturates at about 100 systems for S.D.-aggregated = 0.45 • S.D.-individual.

For a geographical dispersion radius > 500 km, SCC ~0.05, it takes only 5 sites of roughly the same capacity, to reduce S.D.-aggregated for the same individual intra-day and inter-day variability to +/- 2.5% and +/-5% respectively. Increasing the number of sites has diminishing returns, so the spatial smoothing for SCC ~0.05 starts to saturate near S.D.-aggregated = 0.22 • S.D.-individual. For 50 systems over a 500km radius area, the intra-day and inter-day aggregate variability falls to +/- 1.3% and +/-2.6% respectively.

Fig 10: spatial filtering of clear-sky irradiance fluctuation

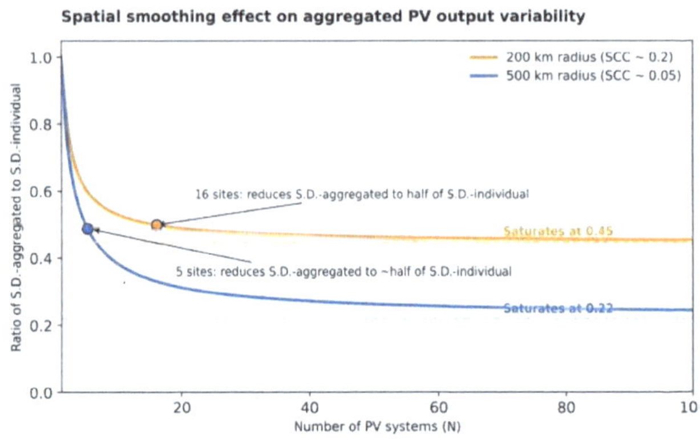

Flexible Dispatch

The standard grid frequency tolerance is +/-0.5%. Hence, the power variability of the aggregate PV dispatch has to be reduced below this tolerance limit. The solution is to combine spatial distribution with flexible dispatch from DC:AC oversizing (Ch10). The practice of DC:AC oversizing is routine for PV projects. We shall take the case of 16 systems geographically spread over 200km, with a SCC of 0.2, to illustrate how the PV dispatch variability can be reduced below the grid tolerance limit.

For the more demanding requirement of inter-day variability, we set a target of preventing a three standard deviations (3 • S.D-aggregated) shortfall event - i.e. a drop of 15% that breaches the grid tolerance of 0.5%. We seek to reduce the aggregated +/- 5% inter-day variability of 16 systems spread over a 200km radius area to less than +/- 0.5%. The probability of aggregated power falling below three standard deviations is ~0.135%. Excess power is not an issue as all PV systems come with built-in downward dispatch. Oversizing is required to reduce power output variability shortfalls of -15% to less than -0.5%. Assume $1MW_p$-ac supply is expected from each system. The occurrence of a 3 • S.D. shortfall event (0.135% probability) would incur a loss of $2.4MW_p$ (15% of $16MW_p$) for its duration (below the grid expectation of $15.92MW_p$ supply, i.e. 0.5% less than expected $16MW_p$). Oversizing, across all 16 systems, for a headroom of extra $2.8MW_p$ capacity, i.e. a 1.175:1 DC:AC ratio each, or 10 oversized at 1.28: 1 each, would take care of the extreme shortfall event for the grid load balancing area.

Figure 11: Oversized DC/AC portfolio for headroom reserve

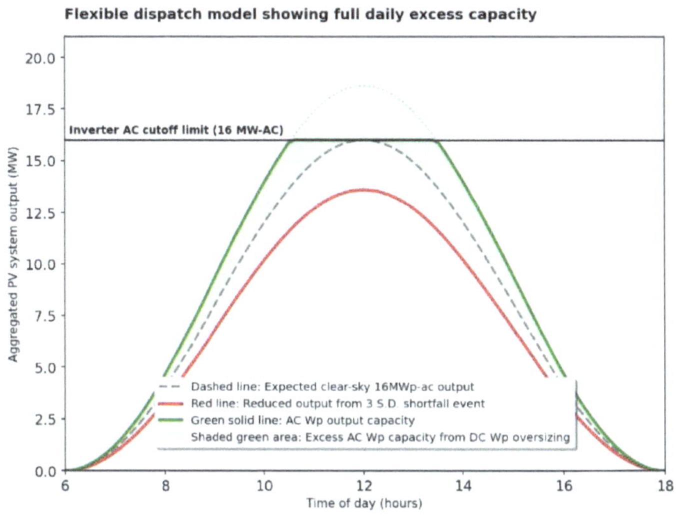

Real Options Analysis (ROA)

PV systems have mainly been deployed for fixed dispatch PPAs, without any requirement of flexible dispatch headroom reserves. One reason is the lack of financial incentive for investing in the reserve capacity. There is no technical impediment to DC:AC oversizing for headroom reserves. Indeed, DC:AC oversizing is fairly common for fixed dispatch contracts.

The missing link is a mutually beneficial financial arrangement for holding the oversized DC capacity in reserve. The grid utility has to financially value the reserves in an appropriate manner – the right to call upon the reserves, only when needed.

ROA provides a mature financial valuation method for the purpose. The option to call upon the oversized PV system is a real option, a Call Option, with a strike quantity (instead of a strike price): the option is exercised by the grid utility only when the strike quantity is exceeded (there is a shortfall beyond grid tolerance).

In the previous section, we showed how oversized PV capacity can be tapped on demand to cover for unpredictable clear-sky irradiance fluctuation. This oversized PV capacity can also be called upon for overall gaps in supply and demand forecasting on the grid.

To illustrate how it would work, we assume the sixteen oversized PV systems are connected to a 100MW-ac load balancing area with an average demand forecasting error of +/- 3MW-ac, including the PV and other power supply variability, during the daytime hours and a grid tolerance of 0.5%. Spot prices for power supply from other sources, called peakers, to meet excess demand are on average \$200/MWh. Oversized PV systems can compete with the peakers by offering their headroom reserve output at \$100/MWh.

The PV flexible headroom output will be maximum for the peak hour. We demonstrate the ROA calculation for the peak hour, with an adjustment for five other daytime hours based on an approximate PV output profile for each hour. Our purpose is to illustrate the ROA principle - an exact calculation can be done for fully simulated PV output at the given location (hourly values for the thirty years duration in standard software e.g. PVSYST).

E (dispatch) is the expected dispatch per peak hour for the grid demand shortfall (E stands for expected):

$$E\ (\text{dispatch}) = E\ [\min\ (\max\ (E_t - K, 0), R)]$$

E_t is the net grid supply vs demand deviation at the time with an aggregated S.D. of 3MW, K = 0.5MW (grid tolerance strike quantity: 0.5% of 100MW-ac), and R is the yearly average of 2.8MW-ac of oversized PV power available for the peak hour.

A normal distribution function f (E_t) with a mean = 0, aggregated S.D. = 3MW, is used to calculate E (dispatch):

$$E\ (\text{dispatch}) = \int_{K}^{K+R} (E_t - K)\ f(E_t)dE_t + 2.8 \cdot \int_{K+R} f(E_t)dE_t$$

We obtain E (dispatch) = 0.757MW, for the peak hour.

For a total daily dispatch over six hours a day, we make a trapezoidal curve approximation around the peak hour for the other hours of less output for an average dispatch of 2.43MWh/day.

Annual flexible energy dispatch = 887MWh.

Annual value (savings versus peakers) = $88,700.

Payback = 5 years and 18 days.

Oversizing cost for flexibility = $448,000 (2.8MWp • $0.16/$W_p$ for additional modules, mounting, and cables only).

Sellers' unlevered **IRR** over the PV systems lifetime of thirty years, factoring in annual power attenuation, for investing in the **flexible dispatch** (Real Call Option) **= 19%.**

By spatial distribution, solar assets mitigate the production variability of each other. PV systems with flexible dispatch DC:AC oversizing can cover the aggregated unpredictable variability, without any energy storage for most days (except intense rainy seasons), and broadly contribute to excess daytime demand on the grid.

Spare, oversized PV capacity for flexible dispatch can compete with peakers for real-time supply shortfalls. During intense rainy periods, with inter-day variabilities of 40-50%, energy storage or other reserves will be required, but these situations can be addressed by diligent weather forecasting. The flexible dispatch provision reduces the energy storage requirement, and associated cost, for the rainy periods as well.

Spatially distributed PV systems with suitably sized flexible dispatch, more than take care of the weather-related intermittency concern. Predictable variability of PV output is not a threat to the grid stability. Unpredictable variability is manageable by geographic distribution and buffers for flexible dispatch. PV project site selection should not be merely based on lowest LCOE calculations, but planned in a system-wide context of aggregated, smoother power delivery. Greater PV penetration of the grid is enabled with financial savings for the grid utility, to mutual advantage.

Fast and Primary Frequency Response

PV can contribute to securing a stable grid frequency, by participating in the market for ancillary services (ancillary services refers to grid support and stabilization functions). Fast frequency response (FFR) is initiated on the sub-second time scale, fully developed in a second to two seconds, and can then last for a few tens of seconds to blend into primary frequency response - the limit of PV systems for frequency regulation. Most frequency disturbances are quenched by this stage.

Grid codes mandate default frequency responses for standard grid-connected, grid following inverters (sometimes referred to as a passive frequency response).

Underfrequency: P is reduced linearly, cut off below a certain level.

Overfrequency: P is curtailed, disconnected above a certain level.

The inverter responds in 100ms-1s to the frequency event, but this precautionary response is different to participation in active frequency regulation. Low frequency ride-through (LFRT) and high frequency ride-through (HFRT) are standard Grid Code requirements.

LFRT: inverters remain connected through an underfrequency disturbance fault, for a prescribed time limit, and are programmed to maintain or increase power injection for frequency support – the purpose is to prevent cascading disconnections leading to frequency collapse. The extent of this frequency support depends on the solar resource availability at the time.

HFRT: inverters stay connected for an overfrequency disturbance, reducing system power output to help bring the frequency down. Grid emulators can check the frequency fault-through responses during commissioning.

FFR, primary frequency response (PFR), and secondary frequency are active, dynamic frequency regulation methods. The PFR range is 2-30s, while secondary frequency response, not within the scope of PV systems, extends from 30s to minutes. PV inverters are quicker at FFR than SGs (Ch3). FFR is triggered within 100ms.

FFR can be programmed in a number of ways:

- Droop control: active power P injection triggered by a change in grid frequency, adjusts proportionally to the change in frequency
- df/dt response: inject P proportional to the rate of change of frequency aka ROCOF (df/dt), the response can be programmed to imitate a SG

- Combination of droop control and df/dt response

PV systems are DC:AC oversized by 5-10% for FFR. PV FFR restricts the "frequency nadir": the lowest point reached by the grid frequency after an instability or disturbance. FFR is crucial for the prevention of frequency collapse, the AC grid's death knell. FFR suppresses frequency transients, while PFR aims for steady-state frequency stability.

PFR also benefits from oversized DC:AC systems - a standby capacity margin of active power P for automatic frequency regulation triggered by fluctuating grid conditions or emergency response on demand. Frequency-watt (F-W) response is an additional technique for PFR.

PFR operational modes include:

F-W control

The watts of active power P are dynamically adjusted according to a pre-defined frequency-watt curve (typically provided by the grid operator, from a grid impact study), with a deadband for small frequency deviations that do not require a response. The response is initiated in 20ms-100ms. Grid operators, with integrated communication capability, may change F-W curve setpoints according to varying grid conditions. During periods of high renewables generation, the grid operator may change the slope of the F-W curve so that P curtailment is sharper for overfrequency events.

Droop control

$P_{out} = P_{set} + K \cdot (f_{nominal} - f_{grid})$,

where P_{out} is the dynamically adjusted power output, P_{set} the normal scheduled power delivery, K a response coefficient in %/Hz, $f_{nominal}$ the nominal grid frequency, and f_{grid} the actual grid frequency.

Synthetic inertia for ROCOF

$P_{out} = P_{set} - J \cdot df/dt$,

where “J” is a virtual inertia coefficient in MW•seconds/Hz (or equivalently MW•s^2) - a higher J is set for grids with low SG inertia.

Hybrid

$P_{out} = P_{set} + K \cdot (f_{nominal} - f_{grid}) - J \cdot df/dt$.

Hybrid tuning combines the droop and synthetic inertia control advantages - faster and longer duration response for more persistent frequency deviations.

Rather than being a liability for AC grid frequency stability, PV provides value-adding techniques for its support. Another control mode of frequency-volt (F-V) regulation is typically employed in mini- or microgrids: frequency and voltage are both changed in independent droop controls to maintain a balanced supply of P and Q. [9,10,11,12]

Dynamic Voltage Regulation

Compensating voltage drops with distance, due to inductive reactance, is a substantial cost for AC grids (Ch3). PV systems connected to the transmission and distribution grids can automatically take care of it, at their point of connection. Voltage stability across the entire grid is strengthened by the presence of PV inverters, enabled for voltage regulation functionality. Automatic voltage regulation (AVR) only works locally - the reactive power for voltage support cannot be remotely delivered over long distances. By increasing the distribution of PV systems across the grid, the breadth of AVR is enhanced.

Operational modes for dynamic voltage regulation:

Q(V) droop control

Adjust the reactive power for a local voltage level target.

$Q(V) = K_{VAr} \cdot (V_{reference} - V_{measured})$,

where K_{VAr} is the droop gain coefficient in VAr per volts, $V_{reference}$ is typically a zonal voltage management target, and $V_{measured}$ is the measured grid voltage.

Q(V) constant voltage

Q is dynamically adjusted for a constant V-reference, using a feedback controller and the proportional-integral-derivative method to calculate the Q requirement (or proportional-integral calculations may suffice, as for the other voltage support operations). For this control mode, a master-slave configuration is essential or PV inverters on the same bus may compete with each other to cause voltage fluctuations.

Q(V) volt-VAr with deadband

Q is adjusted according to volt-VAr curve provided by the grid operator, with a deadband of small voltage deviations for which no response is required. For example, the deadband could be 0.98-1.02 p.u. Frequent, unnecessary Q adjustments can cause unwanted voltage oscillations. Outside the deadband, Q is absorbed or injected according to a characteristic curve.

Q(V) dynamic volt-VAr optimization

The volt-VAr curve setpoints for Q adjustments are dynamically calculated and communicated to the inverter by the grid control system.

Q(V) droop and volt-VAr, with deadband, are the most commonly used modes.

Q(V) hierarchical and coordinated control

A further wrinkle of sophistication is Integrating volt-VAr or droop modes with supervisory commands to fine-tune the control for varying grid conditions.

The responses of local Q(V) controls are coordinated by a central controller.

Q(PF) for a fixed Power Factor (PF) range

Maintains a narrow PF range by adjusting Q relative to P:

$Q(PF) = P \cdot \tan(\cos^{-1} PF)$

Q(V) control is a superior practice to Q(PF) because the Q(V) operational modes are direct and quicker to address voltage fluctuations (dV/dt). A speedy response with minimal delay is particularly important for severe transients and faults.

The equation $V = I \cdot (R \cdot \cos\theta + X \cdot \sin\theta)$, where $\cos\theta = PF$, shows the weakness of a simple PF target for voltage control. V can fluctuate because of deviations in I, R, and or X, though cos θ remains the same.

Simultaneous P, Q support

Prioritizing Q for voltage support and curtailing P for additional Q capacity when necessary. Simultaneous P, Q control is suited to Q(V) operational modes. The inverter can act as an SVG to its full Q capacity by minimizing P.

Volt-Watt (V-W) control

V-W is another voltage regulation method, most useful for microgrids, weak grids, or grids with high PV penetration. Instead of Q, the active power P is adjusted for a pre-defined voltage curve. MPPT is disabled or superseded for the P(V) function. Upper and lower voltage thresholds are defined in per-unit (p.u.) values, where the normalized voltage unit is the ratio of the actual voltage to a nominal base voltage. When the voltage crosses the lower threshold, P is reduced according to a specified ramp in steps. After the upper threshold, P is fully curtailed. The primary purpose of V-W is to prevent overvoltage, curtailing P to restrict voltage rise from the inverters to an acceptable level. This option complements simultaneous P-Q control.

Distinct advantages:

- Transmission grid voltage drops: PV Q(V) droop control removes or reduces the need for capacitor banks and static voltage compensators - expensive infrastructure that may require frequent upgrades

- Weak distribution grids with high X/R ratios, e.g. from long feeders: Q(V), Q(PF), and simultaneous P, Q control eliminates voltage sags and swings, reducing power losses in the distribution network
- Granular voltage control across a larger portion of the grid network: a major reason to financially incentivise distributed solar rooftops including Q(V) or Q(PF) at night
- Dynamic Q injection and absorption, instead of static one-way compensation, that can be updated by remote software commands
- Faster response: IGBT speed vs. magnetic time constant of SG AVR
- Wider voltage control range than SG AVR: scalable by inverter capacity

As for frequency, grid codes prescribe low/high voltage ride-through (LVRT, HVRT) specs for transient faults. Voltage fluctuations happen much faster than frequency changes. PV inverters can automatically initiate voltage control within 100ms to tackle dV/dt, faster than SGs or even batteries. There are numerous cases worldwide of improvements in urban and rural grids chronically suffering from voltage instability problems by the introduction of PV systems with fast AVR.[13,14,15]

Synchronous generator limitations for voltage regulation

SGs are concentrated on the grid network - their Q injection and absorption is therefore available at fewer positions on the grid.

The Q absorption capacity of a SG is slower than PV because of the magnetic loop of excitation - the magnetization takes longer for larger rotor windings. This capacity is also constrained by the under-excitation limit curve of the SG. An under-excitation limiter prevents a loss of excitation that would compromise synchronous stability. It also protects the stator from overheating due to deep, prolonged Q absorption. Protection against loss of excitation is critical, as it is a frequent case of SG failure, but it also restricts the Q absorption of SGs in grid over-voltage crises. This handicap was documented in the final analysis of the Northeast US-Canada grid blackout of 2003.

When electric power demand suddenly drops, P falls, and a constant PF grid strategy exacerbates the problem. SGs may not be able to absorb sufficient Q to balance the power factor. Voltage rises as an outcome, putting the whole grid at risk.

The Iberian grid blackout of 2025 has been characterized as an overvoltage driven event - soon after inverter-based solar and wind power plants were allowed to participate in dynamic voltage regulation.

PV systems deployed on the power network, with Q(V), P(V) enabled, can solve the problem for a grid overly reliant on SGs for voltage stability. Programmed for dynamic voltage regulation, PV electronics can rapidly absorb Q to reduce voltage surges. A more uniform, stable voltage distribution is maintained across the grid network.

PV breakthrough: dynamic voltage support anywhere, anytime

Reactive power (Q) from the PV system, unlike active power dispatch (P), does not depend on the solar resource - only on the capacity of the inverter electronics. Power consumption for Q injection/absorption is insignificant. PV is ubiquitously deployable on the grid. Zonal voltage targets for Q can be tailored for local grid conditions. Grid operators can remotely program Q(V) curves for varying use-cases. Q(V) from PV is thus fast, economical, and adaptable for optimal effect.

Advanced Control Models

Although grid codes prescribe separate responses for frequency and voltage disturbances, the two electrical parameters are fundamentally coupled together in AC grids. Multiplying grid interconnections, varying inertia conditions, an increasing diversity of generators and loads, demands the development of new grid control paradigms, in particular more timely, granular, and simultaneous optimization of frequency and voltage control. Earlier detection of emerging grid instabilities for preventive action, before the situation goes out control. On the flip side, preventing exaggerated responses to grid disturbances.

The development of a new complex frequency metric is noteworthy in this context:[16]

Complex frequency = $(dV/dt)/V + j \cdot \omega$,

where the real part of the complex frequency control variable d(lnV)/dt measures the rate of change of logarithmic voltage, i.e. the ratio of rate of change of voltage to the voltage, and the imaginary part "ω" measures the electrical angular frequency deviation. Voltage dynamics and frequency instability are analytically linked together for joint control. The modulus or argument of complex frequency can provide a more constructive and decisive control signal, than responding to voltage and frequency variations in isolation, considering how the two parameters are fundamentally coupled together. Faster, unified control of voltage and frequency deviations is enabled on the electrical buses of the grid or for device-level monitoring of grid connected generators e.g. the PLL of a PV inverter. This control metric offers potential advantages over the traditional approach of separate frequency and voltage control responses:

- Unified information about the instability severity and cause
- Directing more granular and optimized corrective actions: P, Q combinations for frequency and voltage
- Greater speed and accuracy in the control feedback loop

Innovative control methods are of high importance to keep pace with the rapid development of the modern electric grid, a higher penetration of renewables, blend of DC and AC networks, and sensitivity to faults from multiplying grid interconnections. Advanced control methods, algorithms that take greater advantage of the more granular monitoring and fast feedback possible in power electronics, are the need of the hour.

CF is shown to be a generalized eigenvalue of the grid system matrix, a fundamental physical property of the electric power network. Grid management to date relies on a giant system matrix – covering generators, buses, their parameters and interactions – simplifying a complex web of differential equations. Eigenvalues of this giant system matrix , linearized for small signal deviations, are derived using a number of simplifications. These eigenvalues, natural modes of the power network, are used to plan independent frequency and voltage response protocols of the grid code to stabilize the system. A key premise is that voltage and frequency variations occur at markedly different rates (dV/dt >> df/dt), so that the two can be decoupled for stabilization responses. These simplifications are risky for increasingly sprawling, interconnected grids with more complicated generators and loads. Alternatively, a CF divider formula (CFDF) is applicable to instantly calculate real-time control targets across the grid. This approach, derived from differential topology analysis of the system matrix, enables combined frequency-voltage responses. Control targets from the CFDF can be recursively calculated in real-time for new generators and loads added to the grid. A robust, adaptable grid code is implementable for the challenges of greater renewables integration and heavy-duty loads like data centres.

Clouds

Clouds pose a thorny problem to PV systems. The effects have been thoroughly studied and characterized. Severe cloudy weather and rainy days require the planning of modular energy storage systems (ESS) across the grid. The grid integration methods we have discussed, in particular spatial distribution, play an important role in a mitigation strategy. Total ESS requirement to bring power variability within the grid tolerance for a load balancing area is reduced.

On cloudy days, "flicker" is a symptom of rapid voltage fluctuations. Flicker in lighting is monitored on a P_{st} quality index ("st" stands for short-term flicker severity), with $P_{-st} > 1$ requiring corrective action. Cloud-induced flicker in the range of 0.8 to 1.5 for high PV penetration is a flagged risk. Flicker is not solely caused by the effects of clouds on PV systems, but also occurs in weak grids with high X/R lines (or low short-circuit ratios). Poorly planned PV integration lacking active voltage control exacerbates the problem. We have shown that PV voltage control does not depend on the available solar resource (whether it's cloudy or not), but only on diligent system configuration that can solve the problem of flicker on its own.

Small rooftops appear like points to passing clouds. Complete occlusion can occur with an extreme 90% power drop in a second. The duration of the eclipse depends on cloud velocity - it can pass in half a minute or longer with varied speeds ranging from 2-

10m/s to 15-30m/s in storms, causing power spikes in these time intervals. Spatial decorrelation by geographic distribution over a radius greater than 20km-50km, for a lower SCC, considerably reduces the aggregated power output variability for shared ESS planning. ESS on the load balancing grid network can be blended for the region's cloudy and stormy weather patterns: flywheels to address power spikes in 10-50ms and batteries smoothing power output for the longer duration of several hours. Behind-the-meter battery ESS for solar rooftops is configurable to support the wider grid network against extreme weather events.

For larger ground-mounted systems, the cloud occlusion is slower for gradual power drops from partial shading. There is less vulnerability to sub-second fluctuations, except for edge scattering from the clouds. Edge scattering can induce spikes of higher power output for the grid to absorb. The ramp rates for cloud-driven power reduction in larger ground systems range from 2% to 30% per minute. Distributing these systems, e.g., in a 50-100km radius area sharply reduces the combined ramp rates by as much as a fifth to tenth.

Batteries coupled to larger-scale PV adequately cover cloud effects for both short- and long-duration power predictability. Headroom reserves are scalable by the battery energy storage capacities. As for the smaller rooftops, a spatial spread of ground PV systems in the load balancing area reduces the total ESS deployment cost. Hosting a mix of large and small geographically dispersed PV systems with blended short- and longer-duration ESS portfolio is most effective.[17-21]

Energy Storage Systems (ESS)

Energy storage systems (ESS) span a wide range of operational requirements in different energy and power capacity combinations. Flywheels and supercapacitors have high power capacities but low energy densities. Electrochemical batteries offer a greater variety of power versus energy capacity choices. Battery cells are designable for higher power or energy capacities.

Recently, pumped hydro ESS projects have been announced in Australia with 14-hour 1.9GWh energy storage duration and a 40-year lifetime.[22] Compressed air energy storage (CAES) is another long-duration ESS option. Electrical energy can be stored mechanically (e.g. the flywheels), thermally (e.g., the phase transition materials of molten salts), or electrochemically. PV developers are mostly concerned with stationary electrochemical batteries, the lithium-ion family, the new development of sodium-ion, or the vanadium redox flow variety, as these are modular systems easily incorporated in their projects for different geographies. Several commercially deployed electrochemical ESS projects feature 8-hours long storage duration, at a comparable cost to pumped hydro.

ESS costs are generally quoted in \$/kWh, the energy storage capacity, but the \$/kW, cost of power capacity, has to be also checked for a complete technical rating. The lifecycle economic figure of metric, corresponding to the LCOE of a PV system, is the levelized or lifecycle cost of energy storage (LCOS). The calculation methodology is similar but the data inputs are more complicated, mainly because of how the battery performance degradation rate depends on its use-cases e.g., the charging and discharging cycles per day.

It is worth mentioning that ESS are not only needed for renewables, but more generally. Nuclear power plants also need ESS - critical for emergency power backup to avoid meltdowns. Aside from emergencies, it is difficult to achieve perfect accuracy in demand forecasting and the scheduling of power supply.

Short or long-duration ESS is commonly referred to as a headline system selection choice, but it is really about whether you want high energy or power capacity. A longer duration ESS can fulfil all the functions of a shorter duration system, if it has the same power capacity i.e. ability to transfer energy at the same rate. The ESS unit cost per kWh generally drops for longer duration. Balance of system costs, that of components other than the battery packs, are spread over a greater capacity. Typically, the system is planned for over two hours duration. Shorter duration ESS, with expensive cells specially designed for high power capacities, are chosen for rapid ancillary services.

An overview of prominent commercial types of electrochemical ESS - listing their distinguishing features:

Lithium-ion (Li-ion: LFP, NCA, NMC, other anode-cathode combinations):[23,24]

- Conventional cell design: anode (graphite, silicon blend), cathode (varies for different chemistries), separator (porous polymer), liquid $LiPF_6$ in organic solvent, carbon nanotubes have been introduced in the electrodes to improve performance and stability
- Mature industrial supply chain, production economies of scale and scope (also used in the much larger EV market)
- Service lifetime of 15 years, 80% capacity retention
- Full system costs: energy capacity capex of $55-$120/kWh (full means the whole integrated system including batteries, controller, HVAC, power conditioning), depending on power capacity, storage duration, and scale
- High cell energy densities: volumetric 250-730Wh/L, gravimetric: 150-300Wh/kg (varies by chemistry)
- NCA highest energy density > NMC > LFP
- LFP most common due to safety, lifetime, stable performance
- Round-trip cell efficiency (RTE) > 90%
- Depth of Discharge (DOD) > 90%
- No need to refill or top up the electrolyte
- Self-discharge < 5% for 24 hours, < 5% per month after

- Safety: risk of thermal runaway reactions, mitigated by fire retardants in electrolyte – LFP is the most thermally stable option
- Sensitivity to ambient temperature: operating range of 253K-323K, performance sharply drops below 273K
- Expensive HVAC for narrow operating temperature range
- Precise control is required to avoid overcharging and overheating
- Flammable and toxic ingredients in electrolyte, if it leaks
- Degradation rate increases with faster charging-discharging cycles, depth of discharge, temperature
- LFP cell energy retention capacity of ~80% after 4k-5k charging-discharging cycles, ~70% after 10k cycles (strongly depends on C-rate)
- Self-discharge rate of 1-3% per month
- C-rate (cell current charging-discharging rate relative to capacity): 0.5C-1C, discharging: 1C-5C (higher possible but sharp cost and degradation trade-off)

  C-rate is roughly set by the ratio of energy capacity to power capacity

  The internal electrochemical resistance has to be taken into consideration for full power transfer, as it affects the potential difference driving the transfer
- Design trade-off between energy and power capacity

Sodium-ion (Na-ion):[25]

- Similar cell structure and operating principles as for Li-ion
- ~ $130/kWh approaching, closing the gap with Li-ion
- Wider temperature operating range, superior performance at colder temperatures – lower HVAC cost, improves system-level RTE efficiency
- Prospect of lower cost per kWh for the long-run (vast sodium abundance)
- Cell energy densities: 200-400Wh/L, 100-200Wh/kg
- Cell energy retention capacity of 80% after 2k-4k cycles (a priority for competitive improvement)
- Self-discharge rate of 2-5% per month
- Cell charging rate: 1C-5C, discharging: 1C-8C
- Technology challenge of electrode-electrolyte interphase stability
- Hybrid combination with LFP is gaining commercial traction

Sodium Sulfur (NaS):[26]

- Initial cell design was all solid-state: sodium or sodium alloy anode, molten sulfur cathode, and an alumina ceramic solid-state electrolyte
- Requirement of high operating temperature ~373K-573K: increases risks of corrosion, fire
- Estimated higher cost ~$300/kWh

- Service lifetime of 15 years, with 4,500 charging-discharging cycles
- Cell energy densities: 150-400Wh/L, 150-250Wh/kg

Vanadium Redox Flow (VRB):[27]

- Fundamentally different design of redox (process of oxidation and reduction) flow, from lithium-ion, lead-acid, nickel cadmium, and so on
- Two electrolyte tanks for each electrode: a vanadium based positive and negative electrode on either side, pumps for the flow of the electrolyte into an ion-transfer unit in the middle with an ion-selective membrane to separate the flow of positive and negative electrolytes
- Higher upfront cost ~$150-$300/kWh
- System-level energy densities: 20-60Wh/L, 20-35Wh/kg
- Longer than 8-hour duration, unlimited cycles
- Cell RTE ~ 80-90% (cell to system level efficiency drop is lower than LFP)
- Decoupled power and energy densities: no design trade-off as for metal-ion battery systems - power and energy capacities can be independently varied

  Power capacity: size and number of electrochemical stacks for the redox reaction

  Energy capacity: electrolyte concentration and volume in tanks

- Very safe: the vanadium electrolytes use a water solution, preventing the risks of fires and chemical explosions
- Vanadium is not consumed, the electrolytes are completely recyclable
  (electrolytes are more than 50% of the battery cost)
- Easy, low-cost maintenance
- Lower Levelized Cost of Energy Storage (LCOS) from longer lifetime
- Energy capacity retention > 98% after > 20,000 charging and discharging cycles
- Negligible self-discharge rate
- Typical charging rate: 1C, discharging: 0.5C (higher rates are possible, but requires complex system pumping design)
- Service lifetime, warranties equal to or greater than 20-25 years
- Greater flexibility in independently optimising power and energy capacities
- Unique feature of "overload capacity": to deliver more than twice the rated power for short charging and discharging times of a few seconds
- Ability to start a charging and discharging cycle within 1 millisecond for uninterrupted power supply, to take care of power blackouts

- Rapid response for electric grid stability, "ancillary services" such as voltage sag compensation and frequency regulation
- Careful thermal management is required to maintain the electrolytes temperature between 283K to 313K

There are numerous battery cell technology contenders, in a dynamic field, e.g. an iron-air battery[28]aiming at 100-hour long energy storage duration. Vanadium ion battery, e.g. being developed by Standard Energy in Korea, promises an even higher energy retention capacity than VRB and much faster peak rate of 20C than LFP. Other industrial efforts focus on sub-system and system-level innovations - there are several technical layers for improvisation, to gain a competitive edge. The battery chemistry is but one part. The full, integrated ESS architecture comprises:

Batteries: cells are combined to form modules, and then battery racks are made out of the modules. An alternative cell-to-pack architecture, that is gaining traction, avoids the intermediate integration of cells in modules. Cells -> modules -> battery packs in rack assembles -> placed in container with HVAC (heating, ventilation, and air-conditioning system), FSS (fire suppression system), and auxiliaries (e.g. current, voltage, fluxgate, temperature, humidity sensors, uninterrupted power supply, and other supporting devices).

Battery Management System (BMS): hardware and software control unit, integrated with sensors, to manage the batteries. The BMS has to be developed and implemented for a particular battery type. The BMS is responsible for maintaining the health of the batteries. Sensors can monitor the batteries at the cell, module, and rack (also called stack) levels. Electrochemical Impedance Spectroscopy is a key technique for battery condition monitoring, analysis and timely control. Artificial intelligence (AI) self-learning algorithms are being used with the sensing techniques to maintain battery performance: deliver optimised performance and prevent degradation, failure modes. Even though the hardware might be the same, these self-learning algorithms can make a substantial performance difference.

Power Conversion System (PCS): the PCS is the hardware and software unit for bidirectional DC/AC power conversion – it may also be called the hybrid inverter or storage inverter, typically when in a standalone configuration.

Local Controller: a local hardware and software unit to coordinate the functions of the BMS, HVAC, FSS, and PCS. The local controller is physically located in the same containers as the BMS or with the PCS.

MV Substation of Transformer-Switchgear/RMU (TR-RMU): the electric grid interfacing equipment of the medium voltage (MV) transformer and ring main unit (RMU) of the switchgear. SST integration in the ESS is an attractive option. The PCS, TR-

RMU, hardware are located in the same set of containers, cabled together with the larger set of battery containers.

Energy Management System (EMS): the hardware and software SCADA unit for the ESS and the PV system. EMS provides the user interface for the system user to control and monitor the complete PV-ESS system. The EMS is usually procured from a different vendor than the integrated ESS provider.

ESS integration greatly expands the grid integration prospects and operational modes for PV. The LCOE of ESS has reached grid parity in several places. It is no surprise that in the advanced economies, an estimated 80% of utility-scale PV projects deployed last year are integrated with ESS. PV+ESS is also becoming the norm for smaller-scale projects, with ESS costs sharply dropping across all capacities.

PV+ESS integration is either **DC-coupled or AC-coupled.** DC-coupled PV+ESS is preferred for new projects. There is a single DC/AC conversion step controlled by a hybrid inverter, resulting in lower conversion losses and hence a higher overall system efficiency. The PV array and battery banks are configured behind a hybrid inverter on a shared DC bus. AC-coupled wins for ESS retrofits into existing PV deployment or when multiple generators of different types have to be flexibly scaled and controlled together in a unified system. Retrofitting AC-coupled ESS is quicker and more economical. A battery bank and battery storage inverter are connected to the AC distribution board or other shared AC bus, in parallel to the existing PV system and its grid inverter. Battery storage inverters, unlike a hybrid inverter, do not require the MPPT function (has only ports to connect the battery, none for the PV array). PV output to the grid is not affected, in case the parallel battery section fails. Digital smart meters cue the storage inverter to control charging-discharging cycles and bidirectional power flow. If digital smart meters are not available, a current transformer clamp on the outgoing cable from the distribution board to the grid utility meter signals the battery control.

PV+ESS use-cases include:

Power smoothing

We have discussed how extreme weather variation, in particular cloud effects, is taken care of by ESS for tightly predictable power supply to the grid. Flexible dispatchability is strengthened.

Longer duration power supply

The number of such PV+ESS projects, with ≥ 8 hours duration, under construction is fast multiplying. DC: AC oversizing benefits ESS integration.

Peak load shaving and shifting

Storing solar energy to discharge during peak load, high billing rate hours, for arbitrage with time-of-use billing.

Ancillary services

Frequency and voltage regulation are no longer constrained by the availability of the solar resource at the time. Grid-forming frequency and voltage setting, and black-start capabilities, are established.

ESS contracts are structured in more varied business models than PV systems e.g., toll agreements. In toll agreements, the ESS asset owner leases the system to another party for a fixed payment covering their different use-cases. Asset leasing is also common in PV projects but there is a wider variety of use-cases to manage in ESS.

Autonomous Mini/Microgrids

A logical outcome of distributed energy resources (DERs) more economical than the LCOE of large, centralized power plants is the development of mini- and microgrids.[29,30,31] A mini-grid consists of its own hybrid power system that can autonomously operate for most of the day, if not 24 x 7, with redundant power supply, energy storage, and an ability to control demand response (DR). The mini-grid can work in isolation or in an interoperable mode with the larger grid network. Virtual power plants[32]can be thought of as mini-grids embedded in the grid.

A key enabler of mini-grids is the economy and ubiquitous capability of solar electricity. Almost all modern mini-grids have PV systems. The power electronics controls of PV inverters, fast response times, makes it easier to integrate into hybrid power systems. The coupling with ESS expands the functionality to grid-forming and black-start capabilities, with resilience for extreme weather events. Biomethane from organic waste is a complementary DER, based on mature technology, for combined heat and electric power.

Communities can build their own mini-grids, with greater control over their supply and demand balancing, financial costs of electricity consumption. There are no excessive overheads of distribution grid companies to pay off, or the costs related to extravagant capacity planning, white elephant and vanity projects. Supply and demand forecasts are more accurate for a self-contained group of users or rather prosumers. There are no time delays from capital intensive extensions of transmission and distribution grid lines. Shorter distances for the power transmission and distribution reduces electrical losses. The prospect of bottom-up electricity infrastructure development is particularly attractive for economies that suffer from chronic power outages and load-shedding. Centralized power plants and related grid management bureaucracies have neither cured energy poverty nor delivered reliable, economical electric power - in many places around the world. Despite a poor quality of service, consumers are locked into endless rounds of electricity inflation often because of spiralling fuel prices for the large power plants.

DR is applicable to any load where the power consumption is variable without causing a critical interruption, e.g., in the control of energy guzzling data centres, sewage and waste recycling facilities, or water pumping networks. The time taken for a data centre to respond to an information query is adjustable, resulting in a substantial electric power demand variation, without any severe adverse impact - can be sped up for excess energy, or slowed down for supply shortfalls, to deliver rapid load balancing. A few seconds' adjustment can have a pivotal effect on grid stability. Dynamic DR of selected loads removes the need for drastic load-shedding.

Since PV systems and ESS produce DC power, mini-grids often have a core DC architecture though there is a provision for AC interoperability. With appliances running on DC, there are considerable efficiency savings. There is no need for complicated grid frequency management or added expense of reactive power compensation. The mini-grids are insulated from instabilities and disturbances in the main grid network. These DC islands can act as fire-breaks for faults propagating across the grid. The whole grid is made more secure and resilient. It may even be advantageous to re-think the grid as a constellation of many mini-grids. Embrace a decentralized electricity network, powered by renewable DERs, and foster active consumer participation in its management.

Data Centres

The spectre of rapidly growing data centres is a contemporary concern. Grid to chip, or grid to core, design architectures are receiving increasing attention. Power flow within data centres is controlled by dynamic voltage frequency scaling (DVFS): $C \cdot V^2 \cdot f$, where C is the switching capacitance of the circuit hardware, V the voltage of the power supply, and f the clock frequency of digital data timing. It is essentially asynchronous power flow in a synchronized envelope. DVFS enables demand response (DR) capability - MWs of power consumption can be adjusted in milliseconds for grid load balancing. Voltage and frequency of the electronics can be quickly scaled up/down to make the data centres act as electrical shock absorbers for the grid. Speeding up data centre computations can swiftly absorb excess power supply - digital information becomes an energy buffer system. The V^2 effect of power consumption provides extraordinary DR capability. Adding DR to the grid is most useful for FFR and peak load shaving. The financial value of DR can be conveniently calculated from the same ROA methodology illustrated for flexible solar power dispatch. ROA shows that for higher net variability of the grid, the financial value increases.

Industries like oil and gas require massive data processing in batch workloads, e.g. upstream reservoir modelling and computational fluid dynamics. These tasks become dynamically adjustable to relieve grid stress and reduce operational costs.

New 800V grid to chip/core electrical power architectures fit the mini-grid paradigm. 800V DC is widely regarded as a sweet spot for reducing power losses and hardware costs – particularly well suited to PV plus ESS, EV chargers, and solid-state transformer

(SST) technology. Current reduction at 800V, for the same power delivery, slashes resistive power losses. Higher voltages demand significant hardware cost increases for more stringent electrical safety requirements. Data centres can access more economical power from DER, and participate in flexible power sharing. DER owners, including prosumers, can sell surplus electricity to the data centres.

Advanced methods like CF/CFDF can be implemented in an hierarchical manner to provide fast, accurate, and granular options to stabilize the grid. DVFS V^2 control linked to $(1/V) \cdot dV/dt$ in CF. Data centres acting as an intelligent power load – providing shock absorbers and fire-breaks for faults to enhance grid stability.

Data centres may incentivize a greater number of autonomous mini-grids and virtual power plants. These technical and financial drivers are promising. Instead of the villain portrayal, well planned data centres may accelerate grid modernization to everyone's benefit.

--*--

Ch10 references, annotations, and supplementary reading

1. Australian Electricity Market Operator (AEMO) explanation of merit order dispatch: https://www.aemc.gov.au/energy-system/electricity/electricity-market/how-power-dispatched-across-system
   At the point of consumption, the source of the electricity generation is indistinguishable. Consumers cannot tell whether the kWh being consumed is from a coal or solar power plant – except, of course, indirectly by the effect on their electricity bill.
2. A Method for Clear-Sky Identification and Long-Term Trends Assessment Using Daily Surface Solar Radiation Records, 2022, Martin Wild et al
   https://doi.org/10.1029/2021EA002197
3. Worldwide inter-comparison of clear-sky solar radiation models: Consensus-based review of direct and global irradiance components simulated at the earth surface, Solar Energy 2018, José A. Ruiz-Arias
   https://doi.org/10.1016/j.solener.2018.02.008
4. Internal Variability of All-Sky and Clear-Sky Surface Solar Radiation on Decadal Timescales, JGR Atmospheres, 2022, Boriana Chtirkova et al.
   https://doi.org/10.1029/2021JD036332
5. Clear-sky irradiance predictions for solar resource mapping and large-scale applications: Improved validation methodology and detailed performance analysis of 18 broadband radiative models, Solar Energy, 2012, Charles Gueymard
   https://doi.org/10.1016/j.solener.2011.11.011

6. A New Characterization and Classification Method for Daily Sky Conditions Based on Ground-Based Solar Irradiance Measurement Data, Solar Energy, 2013, Byung O Kang and Kwa-Sur Tam DOI:10.1016/j.solener.2013.04.007
7. On the use of the coefficient of variation to measure spatial and temporal correlation of global solar radiation, Renewable Energy, 2016, Rudy Calif and Ted Soubdhan https://doi.org/10.1016/j.renene.2015.10.049
8. Quantifying PV power Output Variability, Solar Energy, 2010, Thomas Hoff and Richard Perez https://doi.org/10.1016/j.solener.2010.07.003
9. Provision of frequency related services from PV systems, Report IEA-PVPS, T14-17:2024, IEA Task 14 Solar PV in the 100% RES Power System
10. Inverter fast frequency response is a low-cost alternative to system inertia, Electric Power Systems Research, 2023, Yamit Livi and Jay Apt https://doi.org/10.1016/j.epsr.2023.109422
11. Research on active participation of distributed photovoltaics in power grid primary frequency control strategy, Process and Energy System Engineering, 2024, Runtao Yang et al. https://doi.org/10.3389/fenrg.2024.1377351
12. Photovoltaic (PV) Virtual Inertia and Fast Frequency Regulation in High PV Power Grids, Shutang Yang https://arxiv.org/pdf/2010.11340
13. Application of large-scale grid-connected solar photovoltaic system for voltage stability improvement of weak national grids, Nature Scientific Reports, 2021, Adetokun et al. https://www.nature.com/articles/s41598-021-04300-w

14. Distributed Dynamic Grid Support using smart PV inverters during unbalanced grid faults, 2018, Badrul Chowdhury et al. https://ietresearch.onlinelibrary.wiley.com/doi/epdf/10.1049/iet-rpg.2018.5761
15. Evaluation of the voltage support strategies for the low voltage grid connected PV generators, IEEE Energy Conversion Congress and Exposition, 2010, De Souza et al.
DOI: 10.1109/ECCE.2010.5617937
16. The work of Frederico Milano, University College Dublin:
http://faraday1.ucd.ie/archive/papers/freqcomplex.pdf
https://www.ieee.be/node/1029
17. An Assessment of the Influences of Clouds on the Solar Photovoltaic Potential over China, 2023, Yuhui Jiang and Bingqi Yi
https://doi.org/10.3390/rs15010258
18. Cloud Speed Impact on Solar Variability Scaling, Solar Energy, 2013, Mathew Lave and Jan Kleissl
DOI:10.1016/j.solener.2013.01.023
19. High-frequency irradiance fluctuations and geographic smoothing, 2010, Mathew Lave et al.
20. Regime-Dependent Short-Range Solar Forecasting, 2016, T.C. McCandless et al.
https://journals.ametsoc.org/view/journals/apme/55/7/jamc-d-15-0354.1.pdf
21. Cloud Induced Enhancement of Ground Level Solar Radiation, 2013, C. Coimbra et al.
https://ui.adsabs.harvard.edu/abs/2013AGUFM.A13G0318I/abstract
22. Pumped Hydro and long duration battery energy storage projects:

https://reneweconomy.com.au/prices-fall-as-first-pumped-hydro-and-two-eight-hour-battery-projects-win-landmark-storage-tender/

23. Navigating battery choices: A comparative study of lithium iron phosphate and nickel manganese cobalt battery technologies, Future Batteries, 2024, Solomon Evra et al. https://doi.org/10.1016/j.fub.2024.100007
24. NCA batteries overview: https://www.ossila.com/pages/nca-battery
25. Commercialization of Na-ion batteries by CATL: https://sodiumbatteryhub.com/2025/05/15/catl-unveils-sodium-ion-batteries-as-lithium-alternative/
26. Progress in NaS battery technology, BASF and NGK collaboration: https://www.energy-storage.news/next-generation-sodium-sulfur-battery-storage-20-lower-cost-say-basf-and-ngk/
27. Advancements in VRB technology: https://www.numberanalytics.com/blog/advancements-vanadium-redox-battery-technology
28. Form Energy development of the rust battery: https://formenergy.com/technology/battery-technology/
29. A DC Microgrid System for Powering Remote Areas, Energies, 2024, Ardriani et al. https://doi.org/10.3390/en14020493
30. Application of mathematical theory to microgrid design and operation: https://techxplore.com/news/2017-11-framework-stability-microgrids-local-power.html
31. Microgrid Knowledge is a useful resource for market news and technical progress: https://www.microgridknowledge.com/

32. Rocky Mountain Institute case-studies of Virtual Power Plants: https://rmi.org/insight/virtual-power-plant-flipbook/

Ch11. Sustainable Universal Electricity Access

"Uncertainty is the only certainty there is, and knowing how to live with insecurity is the only security" – John Allen Paulos

Despite close to a hundred and fifty years of electrification, electricity generation and consumption across the world is highly skewed. A commercial electric power plant was first set up for Wall Street in 1882: a 600kW steam engine with coal boilers in downtown Manhattan. In 1891, a three-phase 8kV, hundred miles long transmission line from a 220kW hydroelectric dam in Lauffen am Neckar to Frankfurt am Main was demonstrated. Since then, the US has built over 5.5 million miles of medium voltage 1-69kV lines and 600,000 miles of high voltage > 69kV-132kV lines.[1] Similarly, Europe has around 500-600,000km of high voltage transmission lines and 18 million km of distribution grids.[2] In contrast, all of Africa has less than 150,000km high voltage transmission and 750,000km of distribution lines.[3] Global variations in the electricity consumption per person per year are stark: US ~12MWh, Australia ~10.5MWh, Japan ~8MWh, Russia ~7.2MWh, China ~7.1MWh, Europe ~6.1MWh, South America ~2.6MWh, SE Asia has a wide spread of ~1-6MWh across different countries (Singapore being an outlier with ~9MWh, using natural gas pipelines for imports), South Asia ~1MWh, and Africa ~0.63MWh (Libya having the highest at 4.3MWh, down from 6MWh in 2013). The global average is ~3.6MWh per person per year.[4] Two prominent reasons for the disparities are firstly, the cost of and access to fuels for electricity generation, and secondly, the burdensome capital costs and logistical challenges of constructing electrical grids over thousands of miles.

Solar electricity does not require any fuel. Neither does it necessarily require power plants located far away for power transfer over long transmission lines. Geographic dispersal of PV systems is important (Ch10), but it does not involve the large distances of remote hydroelectric, coal, or nuclear plants. Hydroelectric power plants obviously have to be located on top of the water resource. Almost all nuclear plants are built near ample water resources, either the sea or rivers. Brown coal power plants, i.e., lignite, have to be located next to the mine – mine-mouth power plants. Plentiful water supply has to be brought to the mine-mouth, at considerable cost. Anthracite and bituminous coal power plants are located close to the sea or rivers, like nuclear power. We understood the reason for it, in our analysis of how large-scale electricity generating heat engines work: applicable to coal, nuclear, and CCGT (Ch2). Access to water, in particular freshwater, is a critical factor. We can add it to our list of major economic inputs in the electricity generation budget: cost of fuel, the building of lengthy electrical grid lines, and water.

Land access has not been so much of a factor. There are tens of millions of km^2 of unused, non-arable land in the regions of South America, South Asia, and Africa. These vast tracts of unused, uninhabited land have not been useful for electricity generation, or any other economic gain, for over a hundred years. The commercialisation of large-scale PV systems completely changes this picture. All of the land has an electricity generation potential in a range of 100-360kWh/m^2 from the PV process, without the necessity of fuel or ample water resource for the electric kWh production.

Another critical consideration is the materials required for building and operating the electricity generating plant. All materials on the planet are finite. Apart from a little space debris now and then, the planet is a closed system for the exchange of matter. It is an open system for the exchange of energy – in the form of electromagnetic energy. This situation places fundamental natural constraints on the available energy conversion and transfer methods (Ch2). If it weren't for the solar electromagnetic energy replenishment, there would be no life on the planet. It seems rational to prioritize the use of the replenishable resource for electricity generation rather than the consumption of finite materials. It also seems rational to prioritize building the electricity generation systems out of materials that can be reused i.e., recycled. Chemical fuels converted to the entropy of thermal energy can never be re-used. There is no recycling possible for combusted fossil fuels. Metals, on the other hand, are inherently recyclable.

With these broad, basic considerations in mind, we outline the potential of different electricity generation systems for sustainable universal electricity access.

Coal

The single biggest source of global electricity generation is coal. There is something like a trillion tonnes of coal to mine, though in unevenly distributed quantities worldwide and varying in energy content quality. Coal mining is land intensive, water intensive, and far from benign to the land.[5,6,7] It is generally not that healthy for the coal miners either, with well documented hazards like black lung disease.[8] The Haerwusu coal mine pit in China spread over 67.1km^2 is estimated to produce 20 million tonnes/year.[9] At 6.5kWh/kg energy content, 33% average power plant efficiency, 70% capacity utilization factor, that comes to 2.2km^2/TWh per year (the mining land per corresponding TWh electricity generation). Adding the overburden dumping area for excavations (1km^2/TWh), necessary coal ash ponds (0.15km^2/TWh), and plant area (0.05km^2/TWh) the total is ~3.4km^2/TWh or equivalently 0.29TWh/year per km^2. Comparatively, PV systems with single-axis trackers can yield 0.36TWh/year per km^2. One might argue that the Haerwusu land is only useful for coal mining, with no competing need, and that its coal provides power for a longer period round the clock. The land can be rehabilitated later, albeit at a substantial cost.

Wasteland might have no competing need, but scarce water resource does.[10] To assess how much water needs to be withdrawn, and some of it consumed, per TWh - we take the case of 500MW, ultra supercritical coal power plants of 45% efficiency, with standard cooling towers. Nearly 100% of the withdrawal has to be freshwater or the equipment will be corroded and damaged:

Total withdrawal for full capacity ~ 4.4TWh: 5.7-8.8 million m^3/year i.e., 1.3-2.2 million m^3/TWh.

Total consumption (mainly cooling towers, evaporation and blowdown): ~1.7-2.5 million m^3/year i.e., 0.39-0.57 million m^3/TWh.

The water withdrawn from the ecosystem becomes unavailable for other purposes. Assuming the 1TWh is distributed at ~3.6MWh per person, the global average of electricity consumption per capita, it will serve 277,000 people: a withdrawal of 8m^3 per person per year, doubling to 16m^3 per person per year to have an industrialized economy like China. Over a 30-year coal power plant operational period, an estimated ~62m^3 (basic economy) to ~125m^3 (industrialized economy) per person per year will be lost to evaporation.

The global average of renewable freshwater withdrawn per person per year is ~500m^3, adding in non-renewable groundwater withdrawals it is 600m^3. About 70% is for agriculture. If everyone had to rely on coal electricity, it would make a serious dent.

Minimum water for basic hygiene, sanitation, drinking per person per year is ~20-50m^3.[11]

The LCOE of ultra-supercritical projects in China, primarily using their own domestic coal and power plant construction technology, is in the range of $48-$55/MWh.

Nuclear

Nuclear power plants are the most time consuming and capital intensive to build. The US has the world's largest nuclear fleet. Vogtle-3 and Vogtle-4 in Georgia are the first nuclear power plants, commissioned in the US in the last thirty years. These plants had capital budget and time overruns of $21 billion and 6-7 years respectively.[12] France has the highest ratio of nuclear power in its electricity portfolio. The completion of its latest nuclear plant Flamanville-3 was delayed from the expected commissioning date by 12 years, with a €9.9 billion budget overrun.[13] These examples are not isolated but part of a general pattern: historically an average escalation cost of over 100% the initially allocated capital budget.[14]

In addition, the maintenance schedules of ageing power plants in France have faced higher downtimes than what was planned (corrosive parts being reported as a widespread factor, among other issues) – e.g., 910MW Bugey-2 maintenance took 214 days longer than expected. Outages systematically exceeding planned maintenance times contributed to a €10 billion debt burden for

EDF, resulting in the French state having to completely take it over.[15]

China builds its Hualong nuclear power plants on a faster timeline, but only uses nuclear for 5% of its total electricity generation, presumably because of the higher cost involved. Nuclear power plants cannot achieve the high operating temperatures required for ultra supercritical steam cycles to compete with coal thermal efficiencies (Ch2). The highest nuclear plant efficiency to date is ~37% (reported in China and Russia). Water withdrawal and consumption is roughly inversely proportional to thermal efficiency, so the water withdrawal and consumption requirements are ~20% higher than for coal.

The water requirements can disrupt plant operations in hot weather. In 2022, there was a major crisis in France because of water shortages induced by hot weather. This summer, another shortfall in nuclear power was experienced in France and Switzerland (at ambient temperatures of only 298K-301K, due to hotter than normal rivers), though not as severe a crisis as 2022.[16]

Small Modular Reactors (SMR), 300-400MWe nuclear plants, have been aggressively promoted in recent years, but it is well known that nuclear power plants become more expensive with diseconomies of scale.[17] Neutron leakage prevention is more challenging in smaller size reactors. The recent NuScale attempt of building SMR in the USA had to be abandoned.[18] A small nuclear power plant, Phenix 250MWe, a fast breeder design, has operated in France in the past, but suffered from chronic technical difficulties (coolant leaks, turbine failures, and neutron embrittlement). SuperPhenix, a follow-up to Phenix, also suffered complex technical problems - was prematurely shut down in 1998 with a $1.75 billion bill and difficult, technical decommissioning challenges.[19,20]

Although the sizes of nuclear power plants provide greater grid inertia, they cannot ramp up/down quickly like natural gas or even coal (Ch3). This handicap limits their functionality in flexible grid operations. Unexpected, sudden downtime of the larger nuclear plants also places a greater onus on backup reserves e.g., the 2,300MW going offline in the TVA fleet (Tennessee, USA) earlier this year, requiring natural gas and pumped hydro storage to make up the difference. [21] Eskom in South Africa has similarly reported GW-scale unexpected nuclear capacity outages.[22,23] These unplanned outages are on top of the 1-2 months mandatory, planned outages of nuclear plants for refuelling.

It is difficult to imagine nuclear power plants being a globally pervasive option, especially for areas with stressed renewable freshwater resource, because of the financial shocks of capital budget overruns, long-term maintenance issues, unexpected downtimes, lack of flexible compatibility with renewables, and water requirement sensitivity. The decades long decommissioning time, with massive associated costs ($500 million to billion dollars per large reactor), is another difficulty.[24] LCOEs are estimated in the range of €120-€180/MWh. The Flamanville-3 LCOE estimate is €122-176/MWh for an IRR expectation of only 2-7%. Actual, final LCOE for nuclear power plants has a sting in the tail of uncertain decommissioning costs.[25,26,27]

Mining for uranium U_3O_8, commonly called yellowcake, the fuel feedstock for nuclear power plants, is not as land intensive as coal but it can have severer intergenerational effects on the environment. Millions of tonnes of waste, known as tailings, are produced from the mining processes of hydrometallurgical leaching and solvent extraction. Radioactive tailings of the mining retain 85% of the total radioactivity of the ore, with long-living radioactive isotopes e.g. 230-thorium (half-life of 75,000 years) and 226-radium (half-life of 1,600 years). Safe and secure disposal of the radioactive waste is an expensive proposition. Wind and water erosion of the radioactive dumping site invariably occurs, resulting in radioactive dust and radon gas emissions. Rivers and groundwater reserves are contaminated. The eventual cleanup can add up to a large fraction of the financial value of the mined uranium, and sometimes may even exceed the latter.

Natural Gas (fossil fuel)

Natural Gas (NG), along with renewables, has been the fastest growing source of electricity generation in recent years - with the exception of China where coal and renewables dominate. It is largely responsible for the displacement of coal in the US, and most frequently sets the wholesale electricity market clearing price in the US and Europe. All electricity generators are paid this price – hence it is critical to the electricity cost burden on consumers.

Direct NG pipelines, though capital intensive (amortized over long lifetimes of ~50 years), offer high energy transport density without the fuel transport financial and energy costs of coal. For OCGT, there is no steam cycle and hence negligible freshwater required, with air used for the cooling. In rare steam injection gas engines, a modest amount of water is required for the cooling.

CCGT uses recirculating cooling, with estimated water withdrawal and consumption requirements of 500-1,200m^3/GWh and 300-800m^3/GWh respectively. Dry cooling in CCGT can reduce both freshwater withdrawal and consumption requirements to 200-300m^3/GWh, but it comes with a power conversion efficiency penalty and higher per kWh generation cost of 6-12%.[28,29]

Liquefied Natural Gas (LNG) is a much more expensive option than NG. It is about 80-120% costlier per MMBtu, with a knock-on effect of increasing the LCOE by 50-100%. The LNG liquefaction, storage, and transport infrastructure cost is 3-5x that of a direct gas pipeline over the same distance. [30]

LNG is also far more water intensive:[31,32,33]

Withdrawal:

- Gas treatment: 500-800m^3/GWh
- Liquefaction cooling: 10,000-15,000m^3/GWh
- Transport: 200-300m^3/GWh

- Regasification: 15,000-20,000m^3/GWh

Total withdrawal: 25,000-40,000m^3/GWh, 250-800x that of NG

Consumption:

- Gas treatment: 300-500m^3/GWh
- Liquefaction cooling: 400-700m^3/GWh
- Boil-off management: 50-100m^3/GWh
- Regasification: 200-400m^3/GWh

Total consumption: 1,000-1,600m^3/GWh, 20-50x that of NG.

Out of the water withdrawals, the use of seawater is possible for regasification (though that is not always the case).[34] The total necessary freshwater withdrawal and consumption requirement is 2,500-7,000m^3/GWh and 800-1,300m^3/GWh respectively.

The problem with NG for global electricity use is the skewed geographical distribution of its fossil reserves. Most countries are importers of NG, vulnerable to fuel price volatility from supply-side shocks. The LCOE of NG electricity markedly varies by location, whether the NG is being imported by a direct gas pipeline or as LNG - prices may vary beyond these ranges due to extreme market fluctuations:

- Domestic producer: $60-$100/MWh OCGT, $40-$70/MWh CCGT
- Direct gas pipeline: $85-$140/MWh OCGT, $60-$100/MWh CCGT
- LNG: $110-$190/MWh OCGT, $80-$170/MWh CCGT

NG and renewables have enjoyed rapid market growth in tandem, with the flexibility of NG peaker-plants (activated for peak demand or supply shortfall) well suited to fixed quantity renewables PPAs, but it has not always worked to the benefit of consumers. In the US and Europe, where consumers are charged the wholesale market electricity clearance price often set by NG, renewables producers have reaped huge profit margins (the solar PV or wind LCOE being considerably less than NG). The market growth of NG electricity is being aggressively challenged by the combination of renewables plus ESS. Displacing NG offers steadier prices and lower electricity price inflation to consumers.

Renewable Natural Gas (RNG)

The energy generating chemical compound in NG, biogas, and biomethane, i.e., RNG, is methane (CH_4). RNG is a purification of biogas to increase the ratio of the CH4 content. Historically, biogas has been mainly produced, without further large-scale purification. RNG is much more useful and valuable than biogas:

Biogas: 50-75% of CH_4,

Natural Gas from fossil fuel deposits: 93-98% of CH_4.

Biomethane/RNG: 94-99% of CH4.

RNG can be used just like Natural Gas (NG):

- Fed into the same pipelines, infrastructure built for NG
- Produce compressed gas (CNG), Liquefied Natural Gas (LNG) from RNG

- Only a small difference in the energy content between NG and RNG of about 0.3-0.6kWh/m^3, because of trace chemicals in the RNG
- One cubic metre, 1m^3, of RNG can be converted into ~4-6kWh of electric power by OCGT, CCGT (efficiency depends on loading)

Biogas and then purified RNG are made by a biochemical process, from renewable organic feedstock, with possible RNG yields of:[35,36,37]

- Agricultural crops also called energy crops (dry): 60-120m^3/tonne
- Agricultural residue: the waste of agricultural crops: 24-60m^3/tonne
- Livestock manure, bovine (wet): 12-36m^3/tonne
- Food waste (wet): 60-150m^3/tonne
- Municipal waste organic matter (wet), including sewage water: 36-90m^3/tonne
- Industrial waste organic matter (wet), including wastewater: 30-120m^3/tonne

Industrial technology to produce RNG:

The main industrial method for the first step of making biogas is anaerobic digestion. Organic matter in digesters is converted into biogas by bacteria inoculation - it is basically the same process as how gas is created in human or animal digestion. It is a mature, widely used, and economical technology. Each type of organic waste has a different energy content, and requires distinct chemical treatment e.g., for pre-treatment processes, wastewater mix, and carefully chosen combinations that are known to properly work together. The most common method to upgrade biogas to RNG is using membranes to separate chemicals, in an anaerobic membrane bioreactor system.

Growing energy crops for biogas production is not worthwhile - would directly compete with scarcity of arable land and water resources used for food production. It is also not an efficient use of land.

A preferred energy crop for biogas is maize (particularly popular in Britain). Around 12,000m^3 of biogas per hectare (ha) is typically possible from maize, that can be converted to about 45,000kWh of electric power, or 80,000kWh of combined thermal and electric power. By contrast, you can easily get over 1,200,000kWh/ha of solar electricity. So, 15-25x more land is required for energy crops than solar to produce the same amount of kWh. Instead, agricultural crop residue, livestock manure, industrial, municipal, and food waste should be used for the production of RNG.

The EU is aiming at 35-40 billion m^3 of RNG production by 2030, that can generate 210-240 billion kWh per year of electricity. Plans include RNG injection into the gas grid network, as well as its use to make bio-CNG and bio-LNG fuels. The residual material in the digesters is composted and sold as agricultural fertilizer rich in potassium and nitrogen nutrients. [38,39]

The competitive RNG cost strongly depends on the size of the production facilities i.e. economies of scale. The current cost range for RNG electricity generation in the EU is €50/MWh to €100/MWh, for the medium to larger sized projects. These costs can be expected to fall with increasing economies of scale.

For greater than 1200m$^{3/}$hour RNG facilities (corresponding to biogas digesters of greater than 2000m^3/hour yield), a rough break-up of the expected total energy generation cost over a 20-year+ operational lifetime:

- Feedstock: $10-$20/MWh
- Cap Ex: $25-$60/MWh
- Operation: $20-$35/MWh

The feedstock collection cost will depend on the local organisation of waste recovery and planning of the digesters network locations to reduce transport cost. It can also become a negative cost, if collection fees are chargeable for the removal of waste.

The main capital costs are civil works, building the digester, pre-treatment facilities, wastewater facility, purification plant to upgrade biogas to biomethane, and storage of impurities (mostly carbon dioxide). The residual digestate in the digester is valuable fertilizer, rich in nitrogen and potassium, for the agricultural market. Greatest cost reduction from economies of scale lies in the capital costs. The plant capacity utilization is also a significant factor. The main operational costs are the energy input needed to operate the facilities (e.g., for maintaining the required temperature and driving the membrane separation process), membrane replacement, chemical additives, and general maintenance. Cheap solar photovoltaic power can be used to reduce the operational power cost.

RNG substitution of NG imports can be particularly rewarding for countries that rely on expensive imports. Distributed RNG power plants can fit into mini-grids, complementing PV for electricity during the evening hours.

The global electricity generation potential of RNG is largely untapped:[40-45]

- Agricultural energy crops: excluded, to avoid food competition
- Agricultural residue: the waste of agricultural crops: 145-1370TWh/year from 1-3.8 billion tonnes/year
- Livestock manure: 225-825TWh/year from 3.1-3.8 billion tonnes/year
- Food waste: 470-2,250TWh/year from 1.3-2.5 billion tonnes/year
- Municipal waste organic matter, including sewage water: 450-1150TWh/year from 2-2.2 billion tonnes/year
- Industrial waste organic matter, including wastewater: 1380-6,625TWh/year from 7.6-9.2 billion tonnes/year

A possible total of 2,670-12,220TWh/year. For a long time RNG was restricted to small-scale projects, but in recent years 10-100x MWe projects have multiplied, with even the oil and gas majors increasingly involved in developing this renewable electricity resource.[46,47,48]

Hydroelectric

Hydroelectric power, or hydropower for short, is perhaps the oldest form of electricity generation. There are two categories: reservoir-based and run-of-river projects.

Run-of-river extracts energy from the flow of a river by diverting the natural course of the river through turbines, without the use of major water storage, relying on the momentum of the river for its diversion.[49,50] Turbines extract energy not so much from the kinetic energy of the river motion, like in ocean wave motion, but primarily the potential energy from small head differences e.g., from weirs, stepped ponds. The run-of-river power output, P, is given by

$P = \rho \cdot Q \cdot h \cdot g \cdot \text{Eff-turb} \cdot \text{Eff-gen}$, where ρ is the density of water, Q the diverted volumetric flow rate, "h" the height of the head, "g" is acceleration due to gravity, Eff-turb: efficiency of the turbine, and Eff-gen: synchronous generator efficiency (asynchronous induction generators might be used for small projects, typically for off-grid application). The head h is on the order of a few metres to tens of metres. Turbine and generator efficiencies range 85-

95%, but can be affected by the variable flow rate. Volumetric flow rate Q (the fraction of volume from the main river) depends on rainfall, its runoff coefficient (water lost to soil infiltration, evaporation, vegetation), and the rain catchment area feeding the river:

Q = Run-off Coefficient • Rainfall Intensity (m/s) • Catchment Area

No matter how large the catchment area, the electricity generation will vary by the rainfall upon it, following a seasonal pattern. Snowmelt rivers are also heavily influenced by rainfall.

During a rainy season, the unpredictable production variation rises. Inter-seasonal variability can vary by more than a factor of 2x, but it is relatively predictable. The inter-day variability is more unpredictable, easily varying by +/- 20-30% during stormy spells of rainfall. A dry spell in the catchment area can reduce the electricity generation to a trickle.

Sudden natural events can have an even more drastic impact, by totally changing the course of the river. A landslide or other upstream event can leave a run-of-river project high and dry. Flooding can also alter the course of the river. There are quite a few instances of run-of-river projects stranded or badly damaged by such events. Projects on the Sunkoshi River in Nepal, the Assi Ganga in India, the Madeira in Brazil, are among the victims of unpredictable weather events.[51,52]

Reservoir-based dams are meant to provide more certainty to hydropower generation. The fundamental physics of how a reservoir-based dam generates electric power, P, is as before by gravitational potential energy across a height drop:

$P = \rho \cdot Q \cdot h \cdot g \cdot \text{Eff-turb} \cdot \text{Eff-gen} \cdot \text{Eff-pen}$, with ρ the water density, Q the volumetric flow of water released from the reservoir, h the height of the head, Eff-turb: turbine efficiency 90-95%, Eff-gen: synchronous generator efficiency 95-99%, and Eff-pen: the penstock efficiency 95-98% (that of the pipes, gates for controlling the water discharge).

The key differences from run-of-river are that Q the volumetric flow rate can be controlled, subject to the amount of water in the reservoir (in turn depending on rainfall in the regional hydrologic cycle), h can be engineered to greater height, and the penstock can be dynamically adjusted for optimal generator efficiency.

Reservoir-based hydropower dams can have a dual purpose of electric power generation and flood management, but there is a conflict in the two functions. Higher water levels increase electric power capacity, but for flood management a lower water level is needed to leave room for absorbing flood water. A reservoir operating at peak power can get into trouble during sudden flooding. The prudent strategy is to prioritize dams for different purposes.

These hydroelectric generators deliver more predictable power than run-of-river. Greater operational flexibility enables FFR, PFR, frequency restoration response, and automatic voltage control. The grid does not have to rely on fuel-driven SGs for black-start capability and setting the grid frequency. Pumped hydro ESS capability and grid support capabilities are enhanced e.g.:

1800MW, 955m head, Grand'Maison, France:[53]

- Variable speed pumped storage
- Total plant 90% generation efficiency, RTE 78-80%
- Response time of 0-100% load within 3 minutes
- Hydraulic Short-Circuit for frequency control in pumping mode
- 100xMW fast and primary frequency response

715MW, 118m, Itaipu, Brazil/Paraguay:[54]

- 98.6% efficiency
- Direct water cooling eliminates heat exchangers
- Black-start capability, frequency regulation and restoration

900MW, 425m, Nant de Drance, Switzerland:[55]

- Closed-loop pumped hydro ESS of 20GWh, > 80% RTE
- Variable-speed reversible-pump technology
- Response time: 0-900MW in less than five minutes

3GW pumped storage, 380m, Bath County, USA:[56]

- 24GWh ESS, RTE 80%
- Dead stop to full load for one unit in six minutes, all six units in 15 minutes

Sediment bypass systems are now designed or where possible retrofitted for hydroelectric dams, but at substantial additional cost. The Asahi and Miwa dams in Japan are examples of sediment bypass systems successfully working for a long period, close to twenty years now for Miwa and longer for Asahi.

Large reservoir-based designs are not without their problems and limitations. The power capacity is influenced by seasonal weather patterns, with a vulnerability to prolonged periods of drought. Over longer periods, these generators are also heavily dependent upon rainfall in the catchment area. Sedimentation and changing weather conditions can cause accelerated degradation.

Countries that relied on a single or few hydroelectric dams for most of their electricity generation learnt this lesson the hard way, finding themselves in deep economic trouble. Some like Egypt, after the sedimentation of Aswan dam, have pivoted to large-scale solar and wind power projects.

Examples of dams that have become dysfunctional or suffer from persistent underperformance – the bigger they are, the harder they fail:

10,235MW Guri, Venezuela:[57]

- Provided 80% of the country's electricity
- Severe rainfall shortage for over a decade
- 132m head reduced to 92m
- Operating at less than half capacity
- 9 out of 20 generators damaged or idle
- 8 to 12 hours of load-shedding in the country, as a result

750MW, 128m Kariba, Zambia/Zimbabwe:[58,59]

- Drought in 2023, reduced it to ~ 300MW, for a national electricity crisis
- Danger of head reaching minimum level
- Cooling water shortages
  (required for cooling the generator, turbine, bearings)
- Over two months outage in 2022
- Power generation stopped again in 2024, from lack of water
- Zimbabwe has started building a 1GW floating solar project (four phases of 250MW) on the depleted lake

450MW, Sanmenxia, China:[60]

- Sediment accumulation to 62% of reservoir
- Derated from original 1,160MW plan to 450MW capacity
- Stator core, turbines, penstock damage from silt in cooling water

2,080MW, 142m, Hoover Dam, USA:[61]

- Permanent head reduction from droughts
- Generator shutdowns, instabilities
- Renovation to pumped hydro, working in tandem with Nevada PV power, being planned to restore its usefulness

Reservoir-based hydroelectric generation has a major evaporation problem. A reservoir requires a large volume of water to generate electricity, and this water demand is aggravated by evaporation losses. To illustrate the point, consider how much additional volume of water is required for merely 1MWh with a loss of 10m in the head:

Volume for 1MWh = $3.6 \cdot 10^9 J/(1{,}000 kg/m^3 \cdot 9.81 m/s^2 \cdot 100 \cdot 0.9)$, where h the net head is 100m, and 0.9 is the plant efficiency (90%).

Volume for 1MWh with head at 100m =4077m^3, volume at 90m = 4531m^3.

Keep in mind the average annual water consumption of a person for basic domestic needs: ~ 55m^3/year.

A loss of 10m in head, means an additional 454m^3 of water is required for 1MWh electricity generation - over 8x a person's annual water consumption. An acute water shortage also turns into an electricity crisis for the population.

Evaporation accounts for up to 16% of the storage of hydro dam reservoirs in water stressed regions.[62] Studies show a 61% reduction in evaporation by the shading effect of floating PV arrays installed on reservoirs.[63] Additionally, the floating PV arrays help reservoir water quality by reducing the growth of undesirable algae. PV systems not only provide additional electricity generation, but also preserve the dam and its power generation capacity. Unsurprisingly, large floating PV systems have been recently commissioned on dam reservoirs e.g., 145MW on the Cirata reservoir in West Java, Indonesia, 600MW on the Omkareshwar dam in Madhya Pradesh, India, and 69MW on the Lajeado reservoir in Tocantins, Brazil (an installation of bifacial solar panels). This trend is accelerating across all sizes of inland reservoirs, and offshore as well, e.g. for fish farms off the coast in Norway. PV systems on the reservoirs combined with pumped hydro storage and water filtration or desalination systems offer multi-functional power generation.

The LCOE of both reservoir-based hydroelectric and run-of-river is very sensitive to geographical conditions, generally in a range of \$20-\$80/MWh and \$40-\$120/MWh respectively. Run-of-river tends to be more expensive because of greater sensitivity to seasonal rainfall patterns. The LCOE of reservoir hydropower is arguably underestimated, because it does not take into account the opportunity cost of the vast amount of water withdrawal - the value of the water for agriculture, drinking water, municipal supply, and the cost of external impacts on the environment. 10km^3 of water withdrawal for a 1,000MW hydroelectric project could conceivably be valued in billions of dollars, particularly for water stressed regions.

An estimated 1,412GW of hydroelectric projects are operating worldwide, including 189GW of pumped hydro storage. Ever larger projects are being announced, e.g. the 60GW, 300TWh/year project construction just started by China in the Tibetan plateau, though these projects are drawing a mixed reaction - those downstream concerned about their access to water for competing needs. Hydropower is growing, but most economies are diversifying away from excessive reliance on it.[64]

Wind

Wind power along with PV has permanently altered modern electricity infrastructure. The temperature difference between the upper and lower layers of the atmosphere sets up a heat engine with thermal energy gradients, driving wind currents across the planet in a complex pattern, generally from the tropics to the poles. The kinetic energy in these wind flows can be extracted by mechanical to electrical energy conversion. A rotor with blades for wide spatial span can be positioned in the air, to tap the kinetic energy. Wind turbines carry three rotor blades (though some may prefer two), for practical engineering compromise involving cost and mechanical stability. The available mechanical power for the rotor span is given by

$P = ½ \cdot \rho \cdot A \cdot v^3 \cdot C_p$, where ρ is the density of air, A the circular rotor catchment area, v is the air-flow velocity to the power cubed, and C_p is a coefficient that factors in the conservation of momentum, a fundamental rule of nature, as well as energy – i.e., it shows how the conservation of momentum places a constraint on the maximum possible power extraction. If all the wind kinetic energy was extracted, air-flow would cease, and with it the process of energy conversion. Some air must be allowed to pass, placing a hard upper limit on the efficiency of the conversion process for any given velocity of incident wind currents. Applying the laws of conservation of energy and momentum, allows us to derive the optimal downstream and upstream wind velocities for maximum power transfer: v/3 downstream, 2v/3 upstream of the turbine, for v, the site wind velocity. For open air-flow, the corresponding C_p limit can be calculated: 59.3%, for idealized assumptions of incompressible air, no irreversible energy losses from friction, no energy losses from swirling flows in the disturbed air behind the turbine, and an infinite number of narrow blades.[65] The key operative term in the wind power generation equation is velocity, and the fact that it is velocity cubed – a doubling of the velocity increases the available power by 8x, and conversely halving the velocity, reduces it by 1/8.

Wind velocity distributions are highly sensitive to geography, so the availability of wind power varies by much more than the solar resource, exacerbated by the cubed dependence on velocity (the PV power conversion process linearly depends on the incident irradiance). Wind velocity generally increases with height above the Earth's surface, in the lower layers of our atmosphere.

There is a hard upper limit on the maximum wind energy that can be extracted by a single wind turbine. There is also a hard upper limit on the maximum wind energy that can be extracted by multiple wind turbines in the same vicinity. The disturbed air-flow behind a turbine is called its "wake". The wake of the wind energy extraction has an impact on the turbines behind it, affecting the available mechanical power for their rotors. The knock-on effect passes from one wind turbine to the next. The air-flow pattern immediately behind a wind turbine becomes more chaotic, smoothing with distance. A rule of thumb is to leave a distance of about six times the rotor span diameter between successive turbines. Given the finiteness of available space and wind zone distributions, this procedure cannot be continued indefinitely. The available wind energy density over the affected area will keep falling. Therefore, there is a global limit to how much wind energy can be extracted. It comes to 148TW, ~1W/m^2. Practically it is about 80TW, because wind turbines cannot be installed everywhere.[66]

Around 1.05TW of wind power capacity has been installed worldwide, over 90% of it inland. A nacelle at the top of the wind turbine consists of a rotor hub and shaft, bearings, to support the wind blades, actuators for blade pitch control, a yaw with electric motors and braking system to adjust the orientation of the nacelle, a gearbox or direct drive that ramps up the blades speed for a generator, cooling system, sensors and power electronics including an AC/DC/AC converter. A doubly-fed induction generator (DFIG) or permanent magnet synchronous generator (PMSG), preferred for offshore wind, is used with a gearbox or direct drive nacelle respectively. AC power cables running through the steel tower of the turbine transfer the generated power to a transformer and switchgear. The blades made out of fiberglass (or carbon fibre) polymer composite start rotating at a specified cut-in speed (as low as 3m/s) and stop at a cut-out speed (as high as 45m/s offshore, most around 25m/s) for safety. Power generation thus does not vary directly proportionally to the velocity-cubed, as in the theoretical formula, but in a more complicated manner.

The maximum power capacity of the wind turbine is rated by its performance in commonly defined measurement conditions analogous to the STC of solar panels. A capacity factor for the wind turbine is defined as the ratio of the actual energy output to the product of its maximum rated power output and the time interval for it. The wind power capacity factor, as for PV, is useful for comparison between different turbines but not with other types of electric power sources. The measurement conditions are arbitrarily selected benchmarks, that do not represent actual operating conditions varying from place to place. For example, if the PV industry suddenly decided to change the commonly agreed irradiance benchmark from 1000W/m^2 to 500W/m^2, the reported capacity factors of all PV systems worldwide would be doubled overnight. It is therefore technically meaningless to compare the capacity factor of a PV system to that of a wind turbine or heat engine power plant. It is more meaningful to talk in terms of energy yields and LCOE, the cost of the energy yield. The positioning of turbines, including their inter-spacing, is planned to maximise energy yield and minimise the LCOE, not to maximise the capacity factor. Wind farm project developers seek to maximise their project income, and are not pushed about achieving arbitrary capacity factors.

Wind farms come in different sizes, power capacities, but are not as ubiquitously deployable as PV systems, nor as modular. Micro-turbines on a house roof or in the backyard will generally have a much higher LCOE (increases by ~10x for <10kW turbines), with performance complicated by the built environment. The techno-economics of wind farms greatly improves with economies of scale. Offshore wind farms are generally more expensive to build than on-shore wind farms (~1.5-2x the cap ex), but generally benefit from higher wind speeds and more predictable power output from more predictable wind patterns. Turbine inter-spacing can be increased offshore to reduce the wake effect, as more space is available. Alternatively, the inter-spacing can also be reduced, if desirable, with less of wake effect penalty due to smoother offshore wind velocity distributions. Wind farm locations are carefully chosen for wind speed thresholds. Onshore and offshore wind farms have reported energy yields per unit area in the range of 1-10MWh/ha per year (20-50MWh/ha turbine footprint) and 2-15MWh/ha per year (40-90MWh/ha turbine footprint) respectively - the reported numbers vary according to how the wind farm boundary is defined. For high velocity wind corridors, wind farms can offer the lowest LCOE among large-scale electricity generation sources, even lower than PV. As with PV projects, the LCOE can be levered by third-party financing so a lower energy yield wind farm may have a lower LCOE. Ultimately, the decision of where to set up a wind farm depends on the profit expectation of the project developer. Typically, wind farm LCOEs span a range of $20-$80/MWh. Older wind farms may have been set up for a higher LCOE when PPAs were available at much higher tariffs.[67-70] Denmark is a world leader in wind power, in terms of the share of wind power in its total electricity mix ~60%, though the country's geography does not have the benefit of the highest wind speeds. China has the greatest capacity of wind power, generating ~1,050TWh from wind in 2024 (for contrast, the total annual electricity generation of Great Britain is 315TWh). Denmark developed its wind portfolio by clever planning on shallow offshore waters, for low construction costs of relatively high rotor hub heights (~110-150m) to access reasonably high wind speeds (~8.5-9.5m/s). The LCOE of Danish wind farms is lower than the global average: ~$45/MWh vs $70/MWh. There is more to optimal wind farm planning than cherry-picking very high velocity wind corridors (very high velocities may frequently exceed the cut-out speed).[71]

Wind turbine sizes are rapidly increasing, with nameplate capacity ratings of single offshore turbine capacities reaching 16-20MW.[72] A 20MW rated offshore turbine in China has a height of 152m from the base of its foundation to the rotor, and blade lengths of 128m. Quality control for larger wind turbines, their stability and durability, in actual operating conditions becomes more critical. If one turbine gets damaged, most often a blade failure, typically the whole wind farm gets shut down till the problem is investigated and remedied. There appears to be an increase in such incidents with the commissioning of bigger turbines.

Hybrid PV-Wind

Hybrid PV-wind power systems have been popular for some time now, particularly when most or much of the wind yield is at night (or more generally, in a complementary weather pattern). Solar panels are conveniently placed in the gaps for the wind turbine wake, for higher energy yields from the same land area. Alternatively, the wind and PV systems are adjacently located to share the grid connection infrastructure, transformers and switchgear, so that the combined LCOE is below that of the sum of each system installed separately.

PV and wind production are advantageously integrated on a shared DC bus for a hybrid inverter feeding the transformer-switchgear. Wind power systems are either DFIG, with a variable and negative slip speed, or PMSG with zero slip speed (a synchronous speed, not to be confused with grid synchronous frequency speed). Slip speed is the difference between the synchronous speed of the magnetic field and that of the mechanical rotation (a negative slip speed for DFIG is necessary to inject power). For both PMSG and DFIG, Q is absorbed or injected by controls independent of the slip speed. Q control is decoupled from that of P.

A common DC bus architecture is preferable for hybrid PV integration in both cases. PV and wind output are optimised by their dedicated power converters - DC/DC for PV and AC/DC rectification plus DC/DC for wind - for the DC bus to the hybrid inverter. The arrangement also suits battery energy storage system integration. These hybrid systems range from mini-grids to large renewable energy parks.[73,74]

Apart from grid-connected projects, hybrid PV + wind is increasingly used for mining operations, with early examples of it in Chile and Australia. Falling battery prices make the growth of PV + wind + ESS projects inevitable, with minimal if any significant use of diesel gen anymore. Mining projects are very sensitive about their operational productivity, improving financial operating margins by using cheaper electricity generation. Oil and gas companies have also been deploying renewables for their offshore projects, to replace more expensive fuel-based electricity.

Hybrid PV-Diesel

Hybrid PV-diesel power systems are also popular, particularly for off-grid applications to save on volatile diesel costs. The diesel generator is typically an asynchronous induction generator with negative slip speed. A hybrid power plant controller is required for optimising the parallel, additive power flows. The frequency of the diesel induction generator is unstable, as its rotational speed varies with the demand load, so the power plant controller ensures it does not severely affect the PLL of the PV inverter. Conversely, it also protects the diesel generator operation from excess PV power supply that can destabilize it through reverse power flow (forcing it to act as a motor rather than a generator). Given the drop in battery energy storage system (ESS) costs, their integration is highly recommended. Diesel and PV power is fed to the ESS for regulation. ESS hybrid inverters deliver stable frequency and voltage for joint operation. The whole system acquires grid-forming capability. Similar design considerations apply to the integration of PV with any fuel-based induction generator in a hybrid power system.

Other Power Sources

Concentrated Solar Power (CSP) using solar thermal power to generate electricity with built-in energy storage has not been as successful as initially anticipated. Competing with the falling prices of PV/wind + ESS, has become more difficult. Part of the reason is CSP relies on direct normal irradiance only, unlike PV, and the optical equipment is expensive – similar reasons to why CPV has not prospered. Geothermal is heavily marketed, and though it may be suitable for niche projects or unique geographies like Iceland, it is unlikely to become a global electricity source - owing to its low energy density ~50-80mW/m^2. Deep boring projects boost the energy density, but add to the cost. Tidal energy has a high power capacity, 3-5kW/m^2 in places, but is limited to a few locations.

The renewable electricity source that might be the most promising but as yet largely untapped is ocean wave power. This resource is essentially the same as wind power, using kinetic energy in the ocean waters. The higher density of water compared to air, and greater predictability of ocean current flow, with accurate forecasts 48-72 hours in advance, makes it a scientifically appealing and promising global prospect. An estimated global potential of 29,500TWh/year,[75] with mean annual power densities in the range of 20-75kW/m. Another advantage is that the ocean wave power resource is widely distributed across the planet. The technical challenge is the development of hydrophobic turbines and generators that work in seawater conditions.

Solar PV in the Big Picture

We have seen that solar PV is not the most land intensive source of electricity generation. The amount of land required for mining quartz for the silicon solar cells is only 0.04-0.16ha/TWh.[76] The deployment area has less of a land footprint than that for coal mining. Uninhabited, non-arable and non-forested land, excluding Antarctica and the Arctic, across the continents is not scarce – adds up to tens of millions of km^2. Ice-free deserts span ~20.3 million km^2.[77] Available land is not a critical bottleneck for scaling electric power supply.

Meaning of Power/Energy Conversion Efficiency

Energy conversion efficiency is a ratio of two quantities: the ratio of the electrical energy output (numerator) to the energy input (denominator), while power conversion efficiency is the ratio of the instantaneous rates of the energy transfer (rate of change of energy output to that of the input). The numerator output is defined and specified in the same way (the measurement of electrical kWh delivered to the grid point of connection or the load), for different energy conversion processes, but that is not the case for the denominator. The nature of the energy input fundamentally varies by the type of electricity generator. Hence, it is misleading to directly compare the energy conversion efficiencies of two generators with fundamentally different types of energy inputs.

Direct comparisons work for energy conversions of the same type. For example, fossil fuels and radioactive fuel use the same process of heat engines to convert the released thermal energy to electrical energy. So, a direct comparison of thermal efficiencies is relevant to that extent. The relative thermal efficiency provides useful information about critical, operational ingredients required for the conversion. The purpose of the efficiency metric is to gauge how efficiently the energy conversion process works with respect to the most substantial and ultimately limiting resource employed for its operation. For fuel-based power plants, the substantial resource is the fuel, and for hydroelectric power, it is the quantity of water per kWh generated.

With the solar photovoltaic process, there is no fuel consumption and water is not an operational input in the kWh generation. The energy input for the denominator is incident solar irradiation. The incident solar irradiation is circumscribed by the land available for the installation of the solar panel collector area (the exposed surface area of the solar panels array, optimally laid out and oriented for the land location and its topography). PV power or energy conversion efficiency is effectively a physical space or land utilization metric – land is the substantial resource. PV systems are benchmarked against each other, by how efficiently they use the land for deployment. Energy yield per unit area is the most important efficiency metric for PV system comparisons. The efficiency of the materials utilization in building the PV system is reflected in the amount of land that is required for the energy yield. A key difference from the fuel-based power plants is that the land is not being permanently consumed (unlike the combusted fuel or fission of radioactive fuel). The land can be repeatedly utilized. Once the useful operational lifetime of one PV system has expired, another PV system can be installed on the same piece of land. The land and its solar resource remain intact.

Another key difference is that for PV the energy input has no transport cost. For fuel-based power plants, the energy input assessment has to factor in the energy consumed for transporting the fuel to the project site. It is for this reason that lignite coal power plants are only feasible as mine-mouth projects (located close to the lignite mine). If the lignite has to be transported long distances, the energy required for transporting the low energy content fuel becomes a significant proportion of the energy input. For techno-economic comparisons between power plants using different types of energy inputs, e.g. renewables versus fuels, LCOE is the relevant metric. Respective energy conversion efficiencies are factored into the LCOE calculation.

Techno-Economic Comparative Analysis

LCOE comparisons have to be modified for the same criteria explained for the project planning of techno-economically optimal PV systems. The recommended formulas from Ch9:

$\sum_{n=1}^{n=30}$ (Consumer Billing Price – LCOE)$_n$ • $E_n/(1+d)^n$, or

$\sum_{n=1}^{n=30}$ (Avoided Electricity Price – LCOE)$_n$ • $E_n/(1+d)^n$

When applying these techno-economic value formulas across different types of electricity generation sources, an additional adjustment is required. The certainty of the LCOE calculation varies for different types of electricity generation sources. While the LCOE for fuel-free sources can be calculated with a high degree of accuracy, that is not the case for fuel-based generation due to the volatility of the fuel cost input over the project duration (the numerator input in LCOE calculation). Therefore, a ratio of the expected returns to the uncertainty in the expected returns is advisable. The importance of this suggestion is underlined by how electricity generation portfolios are locked into severe electricity inflation because of rising fuel prices including extraordinary fuel price shocks. LCOE projections underestimating fuel cost volatility are a recipe for underutilized or stranded assets.

Quality assured PV systems offer the least long-term uncertainty in the projected LCOE calculation. Wind power has a higher inter-annual standard deviation in the available natural resource to convert - En is more uncertain in the LCOE denominator. PV+ESS LCOE is appropriate for a relative risk benchmark.

An analogy for the risk grading of LCOE is the "Sharpe ratio" metric for financial assets - the expected returns for a risky asset compared to a risk-free benchmark, divided by the uncertainty in risky returns. Nothing is completely risk-free so the risk-free benchmark practically means relatively lowest risk.

For portfolio planning, the grid operator allocates capacities for different classifications of roles and redundancies, e.g., a certain capacity is required for FFR versus FRR. Grid-forming PV+ESS is in principle capable of the whole ambit of technical requirements, subject to varying costs for each role. The costs increase for multi-functional roles. ESS is not limited to electrochemical batteries, but also includes other proven technologies like pumped hydro and CAES. Locally available ESS types may vary. For certain roles, other electricity generators may offer a superior, expected techno-economic value to risk ratio.

Land for PV

50,000TWh per year of solar PV, for 1.5-2x the total of global electricity consumption in 2024, is achievable with the latest power conversion efficiencies on ~500,000 km^2, around 2% of uninhabited, non-arable, non-forested land:

12,500m^2 per MW_p and 1.25GWh/MW_p p.a. -> 500,000km^2 for 50,000TWh.

1.12 million km^2 of land is cultivated for biofuel feedstock globally[78] – for much less energy yield per unit area than PV (with biofuels, solar energy is converted to chemical energy in plants and then to thermal energy followed by thermal conversion to electric power). In any case, PV does not have to be installed on out of sight land, unless the local community is so distressed by its aesthetics. It can also be dispersed among agricultural farms for dual-use of the land, located on vacant roofs, inland water reservoirs, and calm offshore waters. Widespread PV deployment might even help make unused barren land more inhabitable, with economic growth opportunities for those living on the fringes of it. There is no toxic air pollution or groundwater contamination associated with PV projects. Indeed, satellite evidence shows that large-scale PV installation on the deserts in China has improved vegetation in these barren areas, and made a positive impact on the welfare of people living there. As PV expanded to 102.5km^2 of the deserts in China, 30.8km^2 of the covered area became greener (clear signs of new vegetation). These findings are aligned to theoretical modelling of positive environmental effects on desert regions. [79,80]

Materials Bottleneck

Silver (Ag) is the scarcest mineral used in PV panels. The amount of silver required for 50,000TWh/year may seem prohibitive at first glance:

6kg Ag/MW_p -> 240,000 tonnes for 50,000TWh; while the presently known global silver reserves for economical extraction are 640,000 tonnes. About 5,380 tonnes of silver are recycled a year.[81,82]

A silver shortage crisis for PV is a faint possibility for two reasons. First, the ongoing substitution of silver by copper, and second, recycling: PV panels are made out of inherently recyclable materials such as glass and metals (silver being the main metal). The EVA polymer encapsulation is recyclable too, but needs shredding. Silver reduction is a priority of new solar cell and module designs, with new designs consuming half the silver per W_p of others. Industrial production of HJT modules has been achieved without any Ag by one major manufacturer and only 3-5mg Ag/W_p by another.[83] Further reductions will follow across the intensely competitive module manufacturing industry – competitors quickly catch up with new advancements.

Water for PV

Freshwater is not a critical requirement for PV electricity production. It is used for cleaning, when needed, but not for the generation of every kWh unit. A wet mop is sufficient to clean the panels, e.g. once every two months. About 0.25litres/m^2 of water is enough for the panel surface cleaning ~100,000m^3-125,000m^3/TW_p, coming to 24-30 million m^3 for the maintenance of 50,000TWh/year - 0.003m^3-0.00375m^3 per year per person globally. Dry robotic piezoelectric cleaning of the solar panels is an alternative method for regions where the water availability is insufficient. Water usage in the silicon supply-chain also deserves attention: for the silicon refining, wafer production, and cell processing. Wafer cleaning is the most water intensive step. For 2,000 tonnes of polysilicon per TW_p, and 250 litres/kg of water usage across the solar silicon supply-chain, it amounts to 20 million m^3 for 50,000TWh or 0.0025m^3 per person globally.[84] The water use for solar silicon is orders of magnitude less than for other electricity generation options with the possible exception of wind power (the wind turbine materials supply chain has to be checked for comparison, particularly the petrochemicals).

Relative freshwater withdrawal requirement for 1TWh/year, versus solar PV:

NG CCGT: ~1000x, excluding water required for NG mining

Coal: ~4,000x

Nuclear: ~5,000x

Hydro: ~400,000x.

Renewable electricity from PV and wind power appears necessary to provide universal energy access, without triggering regional and global water supply shortages.

Heatwaves

All types of power plants, not just renewables, are affected by weather conditions. Unplanned power shortfalls for extreme or unusual weather are the main penalty. Summer heatwaves in Europe for the past few years starkly highlight the freshwater vulnerability of different power plants compared to solar PV. Nuclear power plants, supposedly the bastions of power supply for all seasons, have been the worst affected. Hydro has felt the pinch of water shortages. Heated high-pressure areas have caused wind power stagnation. Thermal limits of conductors for long-distance power transmission are stressed.

What about PV? Hotter summer weather is accompanied by higher irradiance. Hot temperatures reduce solar cell efficiency (V-mpp reduction), but the irradiance boost drives higher I-mpp (more photons to harvest despite less energy conversion per photon). Elevated irradiance dominates for higher PV output in these conditions. Record-breaking PV production is ideally timed for peak air-conditioning hours. There is an evening bonus. The higher output combined with ESS, for excess power shifting, has helped de-stress grids and curtail prices in the early evening hours.

Energy Payback

Another flippant criticism among people radically prejudiced against PV systems is the fiction of a very long energy payback period. There have been bottom-up studies accounting for the energy usage in every solar panel manufacturing step to show that it is not more than a few months (below six months now with more efficient panels), to a year for the less sunny regions. It is not necessary to go through such studies to realize these criticisms do not make much sense. Unsubsidized PPAs are signed at the lowest tariffs for PV, in $/kWh or whatever other currency per kWh, than any other energy source. The PPA is for selling kWh from the solar panels. If more kWh were going into the solar panel production than it was producing, these PPAs would be immediate financial suicide – would never be signed.

Conclusion

We have discussed each electricity generation source in turn, bearing in mind the two main historical obstacles for economical, universal electricity access: the cost of building expansive grid lines and fuel imports. Wind power, though it is fuel-free, cannot be deployed everywhere, thus requiring long-distance grid lines. PV is the most viable fuel-free option for ubiquitous electricity generation, deployable just about anywhere in compact, low-cost grid networks. RNG from organic waste is an attractive supplement, because it doesn't require fuel imports, mitigates organic waste, and is well suited to distributed electricity generation. ESS is crucial for any electricity generation mix on modern grids, in a multi-functional role of growing importance. Hybrid combinations of PV, wind power, RNG, and ESS extend renewable electricity to smooth, reliable 24x7 grid dispatch. PV plus ESS PPAs are covering greater periods of the day, extending into the evening hours. A standalone 5.2GW PV plus 19GWh ESS system for 24x7 grid power supply is under construction in Abu Dhabi. Another example is the Oasis de Atacama 2GW PV + 11GWh ESS in Chile, due for completion soon, signed up in advance for round-the-clock (24x7) fifteen-year PPAs.[85,86] Heat engines, electromechanical generators cannot match the control precision and system response speed, the flexibility of these integrated PV-ESS power plants.

The third obstacle for the long-run, and the toughest constraint, is the energy-water nexus. The hydrologic cycles on the planet are unlikely to dramatically change to increase the water availability per capita for a growing population. Therefore, it seems eminently sensible that global long-range electricity planning prioritizes the generators with lower water intensities. Water desalination might have to be expanded sooner rather than later, but this process is energy intensive itself. Electric power to drive the filtration should be cheap, needless to say, with a minimal water consumption penalty (i.e. water used in the electricity generation) - the benchmark is PV/wind-powered reverse osmosis at lower than 2 -

4kWh/m^3 energy consumption for a cost of less than \$0.04/kWh.[87]

We are not arguing that future electricity generation portfolios should exclusively consist of PV supported by ESS, or expansive grid networks with large, remotely located power plants should be completely relegated to the past. On the other hand, it is clear that conventional power plants, with centralized technical and financial controls, have been unable to deliver equitable global electrification. Fuel-based power plants are inflationary for the long-run, while solar electricity generation is deflationary.

We are certainly not advocating the shutdown of the petrochemicals supply-chain - materials like EVA for solar cell encapsulation, XLPE for power cable dielectrics, serve vital functions (as in the case of materials for manufacturing electric vehicles, batteries, and wind turbines). Unsurprisingly, the petrochemicals supply-chain is rebalancing its investments and refinery capacity expansions for these new energy materials, demonstrated by what is called the Crude-to-Chemicals (C2C) business strategy. The energy transition to renewables is driven by inexorable techno-economic fundamentals, rather than any misplaced hype.

The electrical grid is a powerful social construct for sharing energy, related infrastructure costs, and the dynamic balancing of disparities in supply and demand. One has to be careful it is not misused as a means of restricting people to a few expensive options of large centralized power plants. Policymakers ensure that the interests of electricity sellers do not result in affordable energy deprivation. An alternative bottom-up, community driven approach of expanding the grid is worth considering, with greater reliance on decentralized, distributed electricity generation.

The grid to a certain extent will always rely on different horses for different courses. Operational instabilities and unpredictable shocks are facts of life for the grid. Redundancy at various timescales will always be necessary, for a varied set of responses from addressing fast fluctuations to restoring voltage and frequency stability against persistent disturbances. No single source is capable of providing the full spectrum of grid support services and power supply that is necessary.

The intermittency tag for renewables is misleading. All power sources have elements of predictable and unpredictable variability. The sudden tripping and unexpected downtime of 500MW-1GW coal and nuclear power plants won't stop happening in the future, so that massive backup for the sudden shortfalls is no longer necessary. A power plant that is built 6-7 years after its expected delivery date, incurring expensive backup for the delay, has a long system unavailability deficit with financial repercussions before contributing a single kWh. Electricity sources, prone to long project completion delays, suffer a greater toll of compounded interest rate costs in the construction finance. Elements of unpredictable variability in each type of generation source have to be recognised, characterized, and mitigated.

Although a mix of electricity generation sources is required, appropriate weights have to be allocated in a portfolio developed for sustainable, universal electricity access. The backbone of the future grid must be most economical, scalable, ubiquitously deployable, without the risk of causing a water crisis. PV best fits the bill, closely supported by other renewables and ESS. A greater focus is therefore warranted on how to optimally utilize PV, rather than propagandistic debates about whether PV is good or bad - misinformation that distracts from due attention to proper deployment methodologies. Complementary electricity resources ought to be planned around a greater use of PV in a coordinated, integrated strategy rather than an adversarial approach.

At present, there is considerable room for improvement in the utilization of PV systems - ranging from quality assured, field-tested deployments to advanced operational modes in particular the provision of flexible dispatch. It is the most economical industrial technology we have for global electrification, directly drawing upon the solar energy sustaining life on our planet.

---***---

Ch11 references, annotations, and supplementary reading

1. 2025 US Infrastructure Report Card
   https://infrastructurereportcard.org/cat-item/energy-infrastructure/
2. European electricity association Eurelectric power transmission data:
   https://www.eurelectric.org/policy-areas/power-distribution/
3. Africa Electricity Grids Explorer
   https://africagrid.energydata.info/
   Rather sketchy data is available, with an estimated 600 million people lacking proper access to electricity, grid connections
4. Electricity consumption by country, approximately:
   https://ember-energy.org/latest-insights/global-electricity-review-2024/global-electricity-trends/
5. Identification of the disturbed range of coal mining activities: A new land surface phenology perspective, Ecological Indicators, 2022, Liu et al. https://doi.org/10.1016/j.ecolind.2022.109375
6. The footprint of coal, Global Energy Monitor
   https://www.gem.wiki/The_footprint_of_coal
7. Increase in land disturbed per ton of mined coal in US Appalachian region:
   https://appvoices.org/2018/08/10/new-study-maps-increase-in-land-disturbed-by-coal-mining/
8. Coal Miner's Pneumoconiosis:
   https://www.lung.org/lung-health-diseases/lung-disease-lookup/black-lung
9. Haerwusu Coal Mine
   Development Status of Coal Mining in China, 2021, Liu Xinjie et al. DOI:10.21203/rs.3.rs-154477/v1

10. Power and Water at Risk, Union of Concerned Scientists https://www.ucs.org/sites/default/files/2019-09/power-and-water-at-risk-with-endnotes.pdf
11. Freshwater resource data: https://www.fao.org/land-water/databases-and-software/aquastat/en/ https://ourworldindata.org/water-use-stress
12. Plant Vogtle: The True Cost of Nuclear Power in the USA https://gcvedfund.org/plant-vogtle/
13. Flamanville 3, France, 12 years late, 4x initially planned budget: https://www.powermag.com/flamanville-3-reactor-online-in-france-after-12-year-delay/
14. Beyond Economies of Scale: Learning from construction cost overrun risks and time delays in global energy infrastructure projects, Energy Research and Social Science, 2025, Savacool et al. https://doi.org/10.1016/j.erss.2025.104057
15. World Nuclear Industry Status Report 2024 https://www.worldnuclearreport.org/IMG/pdf/wnisr2024-v4.pdf
16. NYT report, among others, of nuclear plants in France and Switzerland affected by summer heat: https://www.nytimes.com/2025/07/03/climate/extreme-heat-nuclear-shutdowns.html
17. Nuclear waste from small modular reactors, Proceedings of National Academy of Sciences, 2022, Krall et al. doi: 10.1073/pnas.2111833119
18. Heavily subsidized NuScale SMR project failure, accompanied by aborted SMR attempts of Transatomic Power, Mid-American Energy:

https://www.climateandcapitalmedia.com/small-modular-nuclear-reactors-a-history-of-failure/

19. Fast Breeder Reactors in France, Science and Global Security, 2009, Mycle Schneider
DOI: 10.1080/08929880902953013
20. Nuclear Engineering International: End of SuperPhenix confirmed https://www.neimagazine.com/news/end-of-superph-nix-confirmed/
21. TVA, Tennessee, USA, nuclear power plants pulled offline to address malfunctions:
https://www.timesfreepress.com/news/2025/jun/25/tva-pulls-sequoyah-nuclear-reactors-offline/
22. Reuters news on South Africa nuclear power tripping:
https://www.reuters.com/business/energy/south-africas-koeberg-nuclear-plant-unit-shuts-down-after-unplanned-trip-2025-03-02/
23. South Africa load-shedding, a recurring problem, due to offline nuclear and coal power plants:
https://www.greenbuildingafrica.co.za/unplanned-outages-surge-at-eskom-forcing-up-to-stage-3-loadshedding/
24. IAEA Nuclear Decommissioning: Addressing the Past and Ensuring the Future - discusses the decades-long timespans and cumbersome budgets for decommissioning
https://www.iaea.org/bulletin/nuclear-decommissioning-addressing-the-past-and-ensuring-the-future
25. The Prospective of Nuclear Power in China, Sustainability, 2018, Yan Xu, doi:10.3390/su10062086
Compares nuclear and coal project economics.

26. Report by the French Cour de Comptes (Court of Auditors) on the breakeven electricity selling price (effectively the LCOE) for the EDF nuclear power plant:
https://www.ccomptes.fr/sites/default/files/2025-01/20250114-La-filiere-EPR%20-une-dynamique-nouvelle-des-risques-persistants_0.pdf
(Within the range of Lazard estimates cited in Fig 58, World Nuclear Industry Status Report 2024, reference no. 15)
27. High Priced and Dangerous Nuclear Power, DWI Berlin report:
https://doi.org/10.18723/diw_dwr:2019-30-1
28. How it Works: Water for Power Plant Cooling, 2013, Union of Concerned Scientists
https://www.ucs.org/resources/water-power-plant-cooling
29. Dry Cooling Retrofits at Existing Fossil Fuel-Fired Power Plants in a Water-Stressed Region: Trade-offs in Water Savings, Cost, and Capacity Shortfalls
https://www.osti.gov/servlets/purl/1882420
30. Natural gas and CO2 price variation: impact on the relative cost-efficiency of LNG and pipelines, International Journal of Environmental Studies, Marte Ulvestad and Indra Overland
This article develops a formal model for comparing the cost structure of the two main transport options for natural gas: liquefied natural gas (LNG) and direct gas pipelines
doi: 10.1080/00207233.2012.677581
31. Sustainability and environmental impact in the LNG value chain: Current trends and future opportunities, World Journal of Advanced Research and Reviews, 2024, P. Simpa et al.
https://doi.org/10.30574/wjarr.2024.22.2.1399

32. Discussion on the Water Intake and Drainage Schemes of LNG Terminal Using the Thermal Drainage from Power Plant, Southern Energy Construction, 2022, Jiang et al. DOI: 10.16516/j.gedi.issn2095-8676.2022.S1.006
33. Modern water-treatment challenges: LNG and other facilities that are expanding due to the shale-gas boom face specific challenges when it comes to ensuring the purity of the inlet and outlet water, Chemical Engineering, Vol 121, Issue 9, 2014, Brad Buecker and Brian Clarke
34. Direct Seawater Cooling in LNG liquefaction plants, Robert Burlingame
35. IEA averages for biogas potential https://www.iea.org/data-and-statistics/charts/average-biogas-production-yield-by-tonne-of-feedstock-type
36. Analysis of various feedstocks used for biogas production, Pote et al. https://www.iea.org/data-and-statistics/charts/average-biogas-production-yield-by-tonne-of-feedstock-type
37. Overview of biogas production from different feedstocks, Boamah et al. https://www.ijsrp.org/research-paper-1217/ijsrp-p7225.pdf
38. European Biogas Association has extensive information on EU plans, including the Biomethaverse project aiming to reduce 2023 production costs of €55-110/MWh by 44%: https://www.europeanbiogas.eu/new-eu-funded-project-is-expected-to-increase-european-biomethane-potential-by-66/
39. Gas for Climate on EU biomethane potential: https://gasforclimate2050.eu/biomethane/
40. FAO Statistics: https://www.fao.org/faostat/en/#data

41. EPA AgSTAR: Biogas Recovery in the Agriculture Sector
https://www.epa.gov/agstar
42. UNEP Food Waste Index Report
https://www.unep.org/resources/report/unep-food-waste-index-report-2021
43. UNEP Global Waste Management Index
https://www.unep.org/resources/global-waste-management-outlook-2024
44. IEA Bioenergy Report 2023
https://www.ieabioenergyreview.org/
45. IRENA Bioenergy reports for different regions
https://www.irena.org/publications
46. US Environmental Protection Agency overview of biomethane:
https://www.epa.gov/lmop/renewable-natural-gas
47. RNG Coalition in the US
https://www.rngcoalition.com/
48. Total Energies expansion in RNG:
https://corporate.totalenergies.us/news/totalenergies-and-vanguard-renewables-start-construction-new-farm-based-renewable-natural-gas
https://totalenergies.com/news/press-releases/biogas-france-totalenergies-starts-its-2nd-largest-unit-normandy
49. Streamflow variability and optimal capacity of run-of-river hydropower plants, Advancing Earth and Space Sciences, 2012, S. Basso et al.
https://doi.org/10.1029/2012WR012017
50. LCOE-Based Optimization for the Design of Small Run-of-River Hydropower Plants, Energies, 2022, Amougou et al.
10.3390/en15207507

51. Ruin-of-the-rivers? A global review of run-of-the-river dams, Environmental Management, 2025, Ian G. Baird https://doi.org/10.1007/s00267-024-02062-5
52. Sedimentation from flooding, affecting run-of-river project: https://www.hydropower.org/sediment-management-case-studies/brazil-jirau
53. https://www.nsenergybusiness.com/projects/grand-maison-hydroelectric-power-plant/
54. https://itaipu.energy/
Itaipu is responsible for 90% of the electricity consumed in Paraguay.
55. https://www.nant-de-drance.ch/en/the-power-plant
56. https://www.dominionenergy.com/about/making-energy/hydroelectric-power-facilities/bath-county-pumped-storage-station
57. Contribution of Guri hydropower shortfall to Venezuela electricity crisis: https://www.vox.com/2016/3/17/11254860/venezuela-electricity-crisis
58. Repercussions of heavily relying on declining hydropower dams: https://www.downtoearth.org.in/africa/climate-change-zambia-to-shut-down-hydropower-plant-as-kariba-dries-up
59. Zimbabwe secures African Development Bank funding for floating solar project on Kariba dam reservoir: https://www.power-technology.com/news/zimbabwe-kariba-dam-floating-solar/
60. Sedimentation problems and management strategies of Sanmenxia Reservoir, Yellow River, China, Advancing Earth and Space Sciences, 2005, Guangqian Wang et al.

https://doi.org/10.1029/2004WR003919

61. Proposal of $3 billion project to store solar electricity in Hoover dam: https://thenevadaindependent.com/article/regulatory-hurdles-await-los-angeles-utility-proposal-to-store-nevada-solar-with-the-hoover-dam
62. Quantifying water evaporation from large reservoirs: Implications for water management in water-stressed regions, Environmental Research, 2024, Hannes Nevermann, et al. https://doi.org/10.1016/j.envres.2024.119860
63. Effects of a Floating Photovoltaic System on the Water Evaporation Rate in the Passaúna Reservoir, Brazil, Energies, 2022, Fernando Roberto dos Santos et al. https://doi.org/10.3390/en15176274
64. International Hydropower Association maintains updated project stats https://www.hydropower.org/news/flagship-2025-world-hydropower-outlook-out-now
65. Derivation of the maximum C_p limit – named after Betz: https://www3.nd.edu/~tcorke/w.windturbinecourse/Aerodynamics_Presentation.pdf
66. Global potential for wind-generated electricity, Lu et al. https://www.pnas.org/doi/pdf/10.1073/pnas.0904101106
67. Global Wind Energy Council https://www.gwec.net/
68. Fraunhofer research on wind power https://www.iwes.fraunhofer.de/en.html
69. NREL research on wind power https://www.nrel.gov/wind/publications

70. IEA wind technology collaboration programme https://iea-wind.org/
71. Denmark state of green wind energy information https://stateofgreen.com/en/publications/wind-energy-powering-the-future/
72. Mingyang 20MW wind turbine in South China Sea: https://www.offshorewind.biz/2024/08/29/mingyangs-20-mw-offshore-wind-turbine-stands-complete/
73. Hybrid wind-solar mini-grid for rural area in Nepal: https://www.adb.org/news/nepals-largest-wind-solar-hybrid-power-system-switched-connect-small-village-world
74. A 210MW wind and 107MW PV renewable energy park in South Australia: https://www.iberdrola.com/about-us/what-we-do/onshore-wind-energy/port-augusta-project
75. Ocean Energy Europe overview of different wave energy technologies: https://www.oceanenergy-europe.eu/ocean-energy/wave-energy/
76. Silicon to Solar Study Overview Report, by Australian PV Institute https://apvi.org.au/wp-content/uploads/2024/02/Silicon-to-Solar-Study-Overview-Report_opt.pdf
77. Solar photovoltaic program helps turn deserts green in China: Evidence from satellite monitoring, Journal of Environment, 2022, Zilong Xia et al. https://doi.org/10.1016/j.jenvman.2022.116338
78. OECD-FAO assessment of cropland cultivated for biofuels: https://www.oecd.org/en/publications/2025/07/oecd-fao-agricultural-outlook-2025-2034_3eb15914/full-report/biofuels_6d29857a.html

79. Climate model shows large-scale wind and solar farms in the Sahara increase rain and vegetation, Science, 2018, Yan Li et al. DOI: 10.1126/science.aar562

80. Ice-free deserts – locations and areas of major deserts: https://en.wikipedia.org/wiki/List_of_deserts_by_area

81. Most recent USGS identification of silver reserves for extraction: https://pubs.usgs.gov/periodicals/mcs2025/mcs2025-silver.pdf

82. Silver recycling per year, 2021 figures: 5,380 tonnes https://pubs.usgs.gov/myb/vol1/2021/myb1-2021-silver.pdf

83. Progress in industrial production of HJT modules with reduced Ag consumption, led by Risen Energy: https://taiyangnews.info/markets/china-solar-pv-news-snippets-july-2-2025

84. Analogical environmental cost assessment of silicon flows used in solar panels by the US and China, Nature, 2025, Saeed Golroudbary et al.
https://www.nature.com/articles/s41598-024-60270-9
Water usage for solar silicon deduced from these findings is ~100-250L/kg. I have taken the upper end estimate.

85. Launch of Abu Dhabi 5.2GW PV + 19GWh ESS power plant construction
https://masdar.ae/en/news/newsroom/uae-president-witnesses-launch-of-worlds-first-24-7-solar-pv-battery-storage

86. Oasis de Atacama PV + ESS project for round-the-clock power supply in Chile – 24 x 7, 15-year PPAs signed with Empresa Minera de Atacama SA:
https://grenergy.eu/projects/oasis-de-atacama/

87. Renewable electricity for water desalination:

Reverse osmosis (RO) membrane desalination, driven by wind and solar photovoltaic (PV) energy, Renewable and Sustainable Energy Reviews, 2019, Mohamed T. Mito et al.
https://doi.org/10.1016/j.rser.2019.06.008

Principal Takeaways of each Chapter

Ch1

Solar dependence

Sun is the only source of continuous energy replenishment for planet Earth.

Electromagnetic energy delivery

Solar energy arrives at our planet exclusively in the form of electromagnetic radiation.

Conservation law of total energy

We can convert, store, and transfer energy in different forms, but cannot create or destroy energy - only change the form of its availability.

Electricity

Electric and magnetic fields guide electromagnetic energy flow in an electrical circuit from the generation source to the load.

Ch2

Entropy constraint

Energy conversions and transfers are constrained by the fundamental, universal phenomenon of entropy.

Natural grading of energy

Different forms of energy come in different grades of entropy. The direction of energy conversion matters; it is easier to convert a low-entropy form of energy to a higher-entropy form than vice versa. Exergy is a related concept of energy quality: entropy creation is exergy destruction.

Thermal obstacle

Thermal is generally the highest entropy form of energy. It is a residual in energy conversion and transfer processes, limiting the efficiency of the processes. Minimizing this residual or waste energy is high priority.

Reliance on thermal energy for electricity

We have to date mostly relied on converting thermal energy to electricity through heat engines: thermal to mechanical to electrical energy.

Fundamental limit on efficiency of thermal to electrical energy conversion

Most of the thermal to mechanical conversion is done by vapour expansion (coal and nuclear power plants) or gas pressure cycles, subject to a fundamental thermal efficiency limit that cannot be exceeded by any means.

Water electricity nexus

Freshwater steam is a critical working fluid for the vapour expansion cycle that has dominated our electricity generation. Steam at elevated temperatures and pressures does the thermal energy transfer for the efficiency limiting step.

Ch3

Universal energy access from electricity

Electrical energy can be delivered at greater speed (close to the speed of light) with lower losses over longer distances, and is more precisely controllable across vast power networks, than any other form of energy that we have harnessed. Electrical power transmission does not slow down with distance. The interplay of electric and magnetic fields provides unique control capabilities.

AC grids created by synchronous generators

Synchronous generators of electromagnetic rotating machines convert the mechanical energy of heat engines running on coal, nuclear, or gas, to electricity. Since both mechanical and electrical energy are low-entropy forms, the engineering design of this conversion step achieves 98-99% efficiency. These machines are naturally suited to AC power - hence most of the world's electricity networks run on AC. Maintaining a single frequency across thousands of miles of grid networks and managing the effects of reactance (e.g., on voltage stability) are challenges for AC.

Grid stability and inertia

Synchronous generators are designed to deliver electrical energy, with independent controls of voltage and frequency to stabilize the grid. Frequency is controlled by spinning turbines, while voltage is typically controlled by DC excitation current of magnetized rotor windings. The massive sizes of synchronous generators provide physical inertia to the grid, slowing down any grid disturbance. But inertia is insufficient to prevent or cope with all forms of voltage and frequency disturbances – and it can also be counterproductive in certain scenarios. Different types of generators are called into action for different types of grid management responses. Grid blackouts and power outages occur due to various causes, including inter-area oscillations, sub-synchronous oscillations, and voltage swings. Frequency and voltage are independently controlled but the two operational parameters are physically coupled so both collapse if one goes out of control.

AC DC mix in modern grids

Modern electric grids are an increasing mix of AC and DC power. DC beats AC power cables for long distance power transmission; the crossover distance is shorter for undersea and underground cables. LV and MV networks host a growing number of non-linear DC loads (e.g., laptops, sensitive electronics, LED lights, data centres) - acting as asynchronous power sinks - as well as DERs producing asynchronous DC power. Grid Codes that evolved for AC power networks have to quickly adapt to this complex new world.

Ch4

Semiconductor link to the Sun

Semiconductor technology, utilizing the silicon p-n junction, allows us to directly and instantly tap electromagnetic radiation from the Sun anywhere on the planet for electricity. This silicon device connection with the Sun enables DERs to produce electricity for the grid at a lower cost than the large synchronous generators.

Continuous technology progress

Solar cell designs have relentlessly improved for higher power conversion efficiencies at a lower production cost. A historical misconception that higher efficiency solar cells would prove too costly has been shredded by these designs. The performance to production cost ratio keeps improving with novel design and industrial manufacturing innovation.

Ch5

Success of cell-to-module (CTM) design and manufacturing

Small solar cells are connected together to form larger electricity generating units of modules. Counterintuitively, among other progress, the solar industry is manufacturing modules that produce more energy per unit area than the sum of their constituent cells - CTM ratios of over 100%.

Field-testing of actual performance and latent defects

PV module performance represented on the datasheet cannot be taken for granted. Even the datasheet performance is based on laboratory test projections of how the module is expected to perform in actual operating conditions. In addition, modules may incur latent defects during manufacturing, transport, and most of all installation. Different field-testing techniques are required to rule out latent defects and establish the actual performance trajectory for the project duration.

Industrial supply-chain

Solar panel enhancements are driven by a diverse supply-chain, including specially designed glass, electrical connectors, integrated electronics, and not least the functional polymers (raw materials sourced from the petrochemicals industry) for encapsulation and backsheet protection. Industrial manufacturing innovations span light trapping to enhance energy yields, materials for longer durability in harsh outdoor conditions, and recyclability of the polymers (the rest of the BOM of silicon, metals, and glass are inherently recyclable).

Ch6

Understanding the brain

Inverters are commonly referred to as the brains of PV systems. PV industry innovation overlaps to mutual benefit with the broader semiconductor supply-chain, particularly in silicon manufacturing processes and power electronics for the inverters. It is important to understand how inverters work in detail to make full use of PV system capabilities, e.g., providing reactive power at night to support the grid, voltage control wherever the systems are deployed, monitoring for safety and early warning signs, and delivering maximum power output from the module strings. Inverter diagnostics and prognostics in the field are as essential as those for modules.

Ch7

Underestimating the mounting design

Mounting structures ultimately determine how and where PV systems can be deployed, not to mention how long they will last. These system components are the most difficult to physically replace - the whole system is uprooted. Yet detailed quality assurance of the mounting design is often overlooked. The choice of the exact steel or aluminium alloy, the detail of structural foundations, and advanced simulations of how the racking will fare in actual operational conditions all have a critical bearing on the fate of the project. Trackers are enhancing energy yields and hence financial returns of ground-mounted projects by double-digit margins. Mounting innovation has extended the reach of PV systems to deployment on water - including large offshore projects.

Ch8

Beware the weakest system link

Considerable attention is paid to purchasing expensive items like the transformer and switchgear, and rightly so, but a cheaper system component often does more damage to the project's fortunes. Poor quality MC4 connectors or poorly crimped connections can bring the whole system down. Cable faults, particularly in the LV DC subsystem, are notoriously difficult to detect - unless one is aware of the diagnostic techniques and test instrumentation for each type of fault.

Ch9

Check it yourself - caveat emptor

Simple equations are detailed for consumers to self-assess PV system sizing for their requirements. The buyer only needs to input data shown on the datasheets of the modules and inverters being offered by the installers - removing the need to access or know how to use complicated, expensive PV software for the evaluation. We also detail how PV software programs work in simulating the expected system performance. Guidelines and selection criteria are provided for procurement of the main system components.

Project planning to common advantage

The levelized cost of electricity (LCOE) of a project is essentially the $/MWh electricity purchasing tariff offered by the project developer, the electric power seller, to the buyer in a power purchasing agreement. We walk the reader step by step through the detailed quantitative process of LCOE calculation, with reference to a realistic, generalized project scenario. Sellers seek to maximise their IRR from the project, while buyers want the lowest LCOE offer possible. Third-party project financiers providing financial capital to the seller are interested in maximising their returns - the cost of capital to the seller. We show how a properly planned project can at the same time deliver a higher IRR to the seller with a lower LCOE for the buyer and higher return to the financier. Project planning is optimised for the financial gains of all three stakeholders, without zero-sum give and take haggling.

Gaming project bidding strategies

Our analysis illustrates how sellers can structure their project planning to offer lower LCOEs than the competition while securing higher IRRs for themselves. The effects of DC to AC system sizing, for the same grid connection power capacity, and project financing nuances for competitive advantage are elucidated.

Lowest LCOE trap

Projects are often awarded on the basis of the lowest LCOE offer, subject to the satisfaction of standard, threshold technical requirements. We numerically demonstrate how awarding a project to the lowest LCOE is financially deceptive and a substantial loss to the buyer in a scenario analysis.

Ch10

Spatial smoothing of temporal variability in the solar resource

Even in clear-sky conditions, atmospheric fluctuations of the solar resource cause unpredictable variability in PV system output. Our quantitative analysis details how careful geographic distribution of PV systems minimizes the variability of the aggregated output within a load balancing area.

Flexible dispatch

DC to AC oversizing of PV systems, in addition to geographic distribution, creates headroom capacity reserves for flexible power dispatch on demand. PV systems are able to meet tight grid tolerances through this arrangement - we illustrate how it works for three-sigma deviations in the solar resource.

Real Options valuation of flexible dispatch

PV systems equipped with DC-to-AC oversized headroom reserves are not only capable of covering solar resource shortfalls for each other, but also participating in real-time peak power supply for excess demand on the grid. The headroom reserves are valued as a real Call Option for the grid operator. We detail the Real Options methodology for financially evaluating this flexible dispatch Call Option - showing how it simultaneously delivers attractive returns to the PV owners and substantial cost savings to grid operators.

Grid stability contribution

A common misconception is that PV systems are a destabilizing factor for the electric grid. The opposite is true for properly planned and operated systems. PV can deliver FFR and PFR (to a lesser extent) for grid stabilization, faster than synchronous generators and even batteries. Automatic voltage regulation is achievable anywhere, anytime (even at night), across the grid network. The various operational modalities for dynamic voltage and frequency regulation are described.

Advanced grid control paradigm

Grid codes provision separate control responses for frequency and voltage disturbances. This historical approach is based on risky simplifications of how AC power networks work. New control metrics like Complex Frequency (CF) enable joint voltage and frequency responses addressing the fundamental physical coupling of the two grid parameters. CF is well suited to implementation in Phase Locked Loops of inverters.

PV+ESS

Most new PV systems are deployed with battery ESS, while older systems are being retrofitted. An overview of the various battery technologies to integrate with PV highlights the pros and cons of each option. PV+ESS is deployable in DC-coupled or AC-coupled configurations, depending on the situation. Longer-duration combinations including round-the-the-clock power supply are an increasing trend.

Ch11

Techno-economic comparison of grid power generators

Routinely quoted metrics such as capacity factors make for misleading comparisons. Alternative techno-economic value formulas are put forward, keeping in view the uncertainties involved in long-range projections.

Future prospects of major electricity sources

Electricity access across the globe is grossly skewed - the main factors responsible for this disparity are identified. The prospects of various electric power generators to deliver sustainable, universal electricity access are objectively discussed in terms of their techno-economic fundamentals, empirical track records, and natural resource requirements. When it comes to land the entire supply-chain of the electricity source is considered, starting from its mining footprint. The water-energy nexus, not land, emerges as the most daunting challenge for global electric power.

--

For any queries, clarifications, corrections, or suggestions for improving the content, please feel free to contact the author Omar S. Cheema at comar@vivantive.com.

---

www.ingramcontent.com/pod-product-compliance
Ingram Content Group UK Ltd.
Pitfield, Milton Keynes, MK11 3LW, UK
UKHW060020051026
15591UKWH00006B/296
* 9 7 8 1 0 3 6 9 3 4 7 1 2 *